CULTIVATING CANNABIS

A Guidebook Intended to
Help Growers from Seed to Harvest

by

Samuel A. Liegel CCA
& Edward A. Liegel CCA, CPAg, CPSS

DORRANCE
PUBLISHING CO
EST. 1920
PITTSBURGH, PENNSYLVANIA 15238

Dorrance Publishing Co
585 Alpha Drive
Pittsburgh, PA 15238
Visit our website at *www.dorrancebookstore.com*

ISBN: 978-1-6376-4313-6
eISBN: 978-1-6376-4627-4

PROLOGUE

Farmers have been cultivating crops since the birth of civilization. It is one of the oldest known professions in existence. For thousands of years, farmers throughout the world have been able to cultivate crops in order to sustain the increased growth of those civilizations.

Times of famine are well-documented throughout history because lack of food was a daily catastrophe for those who were affected. Currently, worldwide access to adequate nutrition is still extremely variable — even with all of the technological advancements in agriculture. Different populations of people have different diets, and those diets vary in cost and availability.

But there is one constant that remains.

People need to eat. It really is as simple as that.

In America, there exists a divide. There's the diet modern farmers are producing at an acceptable price, and the one that is the modern consumer's preference for healthy foods. Time will tell who budges first, the agricultural industry or the consumer, but we believe that the disconnect between the two should be reduced as much as possible.

Consumers would like food transparency to its greatest extent. In the current climate, more than ever before, the modern consumer cares about what foods they put into their bodies. Those goods are intended to be for sustenance, and they should not contain adverse effects to human health.

That is the foundational principle: to sustain life.

Consumers purchase products containing cannabinoids for therapeutic effects and treatments. They do so with the understanding that it will not adversely affect their health or that the positive effects greatly outweigh any adverse effects. Extensive research needs to be established to verify this ideology.

However, one basic principle should be applied as it pertains to these consumers, "Everything in moderation," and that includes cannabinoids.

An entire industry for hemp products has sprung up over the past decade in our country, which in itself is amazing. Many effects from this new industry have already been felt to date. They will continue to happen in the future, and that's not just on a macro- or micro-economic level. This industry has the potential to change our culture. It has the potential to change mindsets, understandings, and how we live our daily lives.

The cannabis plant has many uses. It can be used for therapy, building materials, energy, paper, and composite plastics; this list goes on and on. We need farmers to continually cultivate cannabis plants, providing the raw materials necessary for advancement.

Modern farmers face many obstacles in their paths toward continued success. The agricultural industry carries with it large financial constraints: land values, equipment costs, operational costs, and more. These constraints, along with farmers' abilities to over-produce goods in a free-market economy, create slim profit margins for farmers to survive on.

People continue to migrate toward cities, leaving less and less farmers to carry the load of agricultural production. The time of the generational family farm has passed, and in its wake remains those producers who must continually increase their size, production, or value of goods produced in order to stay profitable.

We have found that most farmers are independent, creative, nurturing, and driven, which are products of the industry that they work in. Most farmers want to take care of their livestock, land, and crops, and they do so in a fashion that maintains the environment we live in. Farmers are also stewards of the land itself. We are only here for a short period of time, and the land that we cultivate or manage will be here long after we are gone.

Farming is a cyclical industry. In the United States, when the stock market is booming, agriculture tends to be depressed. When the overall economy is in recession, the value of the dollar decreases, and agricultural goods flow in greater demand. Times of economic recession are times of increased profitability for farmers. When agriculture is depressed, it affects not only producers' bottom lines, but also it affects their overall mental states.

Agriculture is a dangerous profession, and accidents happen every day. The agricultural industry has a high rate of suicide from farmers who become desperate and believe they have nowhere else to turn. We have clients, neighbors, and friends whose families have been affected by such situations.

So, now we have a "new" crop that sprouted from an emerging industry, which satisfies a need farmers have for new crops with high-revenue potential. That is why many farmers during the past decade have turned to hemp.

We recommend that more farmers continue down this path, while tempering expectations. Growing cannabis will likely not result in rags-to-riches stories for farms. If growers choose to undertake the endeavor, they should do so with the outlook of increased net revenue per acre compared to traditional crops. They should not think they will immediately become rich because that is an unrealistic expectation destined for failure.

Cannabis farming has potential — potential for increased revenue, potential for new products that affect our lives, and potential for change.

Let's set it into motion.

Disclaimer: *The contents of this handbook are intended to be a guide and/or useful reference for both new and experienced growers. Successful cannabis cultivation can be attributed to many variables specific to each grower, and growers will experience many challenges that are out of their control. We believe it's essential to understand which decisions to focus on and which challenges to prioritize in order to increase the likelihood of a successful crop.*

CONTENTS

The guidebook is divided into specific chapters including: feminized cannabis, feminized overview schedule, dual-purpose cannabis, dual-purpose overview schedule, and cannabis approved products with vendors. In-depth details surrounding each topic are discussed with opinions of pros and cons associated with different production techniques. As stated previously, this is not the only way to grow, simply the perspective of agronomists who focused all of their time, professional understanding, planning, scouting, and personal finances to grow cannabis in Wisconsin.

CHAPTER 1

Meet the Growers

Prior to growing cannabis in 2019 in Wisconsin, the authors of this handbook had no experience with cannabis cultivation. However, they collectively have more than 50 years of agricultural agronomy experience (including specialty crops). They have experience in seed sales, soil sampling, nutrient-management planning, scouting, precision agriculture, crop planning, crop budgeting, and crop marketing. They have a family farm, E Legal Cattle LLC, that will soon be a Wisconsin century farm and that boasts multiple generations of farming history. Each generation has had to adapt to the changing agricultural practices of the time to survive. They see modern cannabis cultivation as an extension of that mindset.

Early in July of 2019 representatives from CV Sciences and Bluebird Botanicals flew to Wisconsin to visit our grows and infrastructure. After speaking with them, the growers quickly realized that one successful year of growing experience was a mandate for most large processors.

Those processors found that after the first year, growers made a "year-two jump" and were better-suited to delivering a marketable crop to industry specs. Those processors also mandated they would only consider partnering with growers who had at least one year of experience. We, the growers, now understand many of the challenges that need to be overcome in the first year of cannabis cultivation, and we want to pass many of the trial-by-fire lessons on to our readers.

In 2019, we grew 90 acres of feminized cannabis for CBD production, along with 70 acres of dual-purpose cannabis grown experimentally. We had to overcome many obstacles in order to have successful crops in both models.

This handbook is a collaboration and collective response to the research, greenhouse data, scouting notes, crop reports, application records, and yield data that we attained during the past year. It is by no means the only way to do it and does not contain everything growers need to know in regards to cannabis production. It is meant to be used as a guide or reference for other growers.

We believe that sharing information is the key to success in any industry, and we hoped to have something similar to this available to use this past year. Previously, information within the cannabis industry has not been readily shared by successful growers. We believe that has hindered the industry as a whole. We are Wisconsin Certified Crop Advisors (CCAs) who now have three years of experience in the emerging cannabis industry.

Ideally, this guide will be of use to growers both large and small, who may be inexperienced and experienced, as well as to agriculture professionals who see value in advancing crop planning associated with this crop.

CHAPTER 2

History of Cannabis

Cannabis has a celebrated production-agriculture history not only in the state of Wisconsin but also in the country and worldwide. The plant has the ability to adapt to almost any environment suitable for agricultural production. Perhaps because of that useful quality, it has been cultivated throughout the world for thousands of years. Specific strains, otherwise known as cultivars, have been selectively bred for specific purposes, just like that of other commodity crops.

For example, in the early 1900s cannabis was grown in Wisconsin for its long straight fibers, which were used especially for making strong rope. The progress of strain selection has been halted in the United States for 60 years, which must be taken into account when growing certain modern cultivars outdoors. Specifically, many strains have been bred indoors for decades and are unable to withstand outdoor disease/pest pressures. Currently, cannabis is being cultivated for CBD/alternative cannabinoid production, fiber, grain, and a combination of purposes.

There are many growers who believe they can "grow anything," and for that reason, along with the possibility of increased revenues, they have entered the cannabis-growing industry. Disregarding the motives behind entering the industry, all growers must realize that understanding the management requirements of cannabis will go a long way toward ensuring a marketable crop. This guide was created with the express purpose of delving into those management requirements in order to produce a successful crop.

Growers who have experience with specialty crops, horticulture, greenhouse management, transplantation, or who have understanding of the management requirement of large-scale agricultural crops, are better-suited to enter the cannabis space. Also, those growers who have years of experience with outdoor cannabis on a small scale are better-suited to create crop plans specific to their hemp crops, thus creating increased potential for success.

That's not to say others with less experience can't be just as successful. It all relies on properly managing cannabis during the growing process.

Cannabis is a bioremediator and reacts quickly to a changing environment. This means that the plant will take up and store any and all elements or molecules in excess in its environment. For that reason, understanding the plant's needs, deficiencies, and stresses, while also responding to them in a timely manner, will increase yield.

There is a low probability of success when engaging in the mindset of "plant it, and it will grow." If a field reaches the stage of plant germination and isn't tended or managed, it's unlikely to produce a marketable crop. Certain modern cannabinoid cultivars have been "babied" indoors for decades then crossed with other strains to make them more adaptable for outdoor production.   Others have built-in genetic defenses for specific diseases and environmental stresses. Without knowledge of the genome (identifying cannabis genes associated with natural disease and pest resistance) it's difficult to select genetics specific to environmental pressures in a specific area. All growers should have knowledge of the needs, pressures, and deficiencies associated with growing cannabis in their respective areas; agronomic plans will then need to be adjusted accordingly.

Cannabis in Wisconsin

The cannabis season in Wisconsin runs from April 15 through November 15. Those dates aren't set in stone and can easily vary year to year depending on weather patterns. Seed starts or clones should be started at the beginning of this season. The growing season will occur, and the crop should be in a stored stable state by the end of the season. Processing of the previous years' crops will likely occur year round. Processors will need to adjust their capacities and throughput to handle the coming year's crop.

This guidebook is intended to be a seed-to-harvest reference for growers. It does not contain information regarding processing or marketing of end products.

CHAPTER 3

Genetics Selection

Understand and identify plant types

Feminized cannabis plants can be grouped into three main plant types: sativa, indica, and ruderalis.

Sativas

Sativas grow the tallest of the three, reaching from four to nine feet. They originate from semi-tropical regions of the globe. Sativas have thin leaf blades, which are elongated, and are mid-maturing varietals. Between the nodes on growing branches, sativas tend to exhibit longer spaces than the other plant types, which aids in identifying them.

If planted between May and June they should mature in mid- to late September or early October in Wisconsin. On production acres sativas should be planted in a range of 1,900 to 2,200 plants per acre due to their large size. That will allow for maximum growth potential while avoiding overcrowding issues. Ideally, the plants will canopy by early to mid-July and suppress weed growth.

Indicas

Indicas are smaller than sativas but larger than ruderalis. They range from two to six feet in height and are bushy plants with broad short leaf blades. They are slower growing than sativas, and if planted between May and June they should mature throughout the month of October in Wisconsin.

Indicas originate from northern climates and are said to be "hardier" plant types than sativas. Indicas are short-day maturing. That means they will flower when the hours of daylight become shorter in the fall. Identify indicas by their tendency to have shorter spacing between nodes on growing branches and possibly more nodes within their branching. Due to their moderate size, indicas should be planted in the range of 2,800 to 3,200 plants per acre. Ideally, the plants will canopy by mid- to late July and suppress weed growth.

Ruderalis

Ruderalis plants are the smallest of the plant types, growing from two to four feet in height. The ruderalis gene has some debate to its origins. Most scientists credit the origin to that of Russian descent. Wild ruderalis plants have been identified throughout the Russian countryside.

Ruderalis, or "autoflowers," differ from the other two plant types because they do not flower based on photoperiod. Ruderalis plants will "autoflower" anywhere from three to seven weeks after germination. Because of that, ruderalis plants can be dual-cropped in the same growing season or planted near July in northern climates where summers are short. Ruderalis planted in June in Wisconsin will likely reach maturity in August.

Due to their small overall size, dominant strains of ruderalis tend to limit yields in terms of dry flower pounds per acre (lbs/acre). They can be especially effective at spreading out the harvest season for growers looking to maximize the harvest/drying season. These varietals can be the last ones planted in the spring and the first ready for harvest in mid-August.

Indicas and sativas will not mature until September or October, so having a varietal that is ready in August lengthens the harvest season.

Ruderalis plants will likely never reach canopy unless densely populated (10 to 20K seeds/acre), which equates to large costs to growers. Instead, the smaller plants could be cultivated for weed control/suppression up until the time of intended harvest.

Origin of genetics

Most modern feminized cannabis strains that are bred for CBD/alternative cannabinoid production are a result of crossing genetics from sativas and indicas. Usually, the dominant plant type can be identified in the varietal.

For instance, Trump 2 (T2) or Midwestern genetics are indica-dominant, and Suver Haze or Hawaiian Haze are sativa-dominant.

Understanding the plant type prior to purchase is essential, along with understanding its growth potential. Inquiry and research of plant type via correspondence with the plant breeder should be done prior to cultivation.

Some breeders have reputable Standard Operating Procedures (SOPs) regarding feminization practices and purity of seeds. Purchase of seeds from non-reputable breeders could result in a "mixed bag" of seeds with differing plant types, maturities, males, and even some ruderalis genes mixed in.

An acre that is planted with differing plant types becomes difficult and costly to manage, test, and harvest. Feminization genetics/breeding should be accomplished indoors to control pollen. If attempted outdoors, there is an increased risk of feral or neighboring pollen drifting in and pollinating females that are intended to be producing feminized seed. Outdoor feminization of seed fields tend to result in those "mixed bag" genetics and are extremely costly and difficult to manage during the following production season.

Schedule plantings

Spreading out planting dates of specific varietals throughout the spring can be beneficial in lengthening the harvest window, which maximizes the harvest season.

One strategy (see Figure 1) is to seed the indicas first, which are slow-growing and late-maturing, then seed the sativas, which are fast-growing and early-harvest, and then seed the ruderalis. The late-planted ruderalis (autoflower) will be the first to reach maturity – likely in late July or early August. Harvest can then begin and continue until the second week of November.

Growers who plan to cultivate large acres and need to manage their harvest windows should adopt such practices. This also lengthens the harvest window for processors who need to coordinate harvest times with multiple growers.

Figure 1

Planting:	indica	›	sativa	›	ruderalis (autoflower)
Harvest:	ruderalis	›	sativa	›	indica
	Aug.	›	Sept.	›	Oct.

Select genetics

When selecting specific cannabis varietals for production acres in Wisconsin, there are many identifiable traits to consider. Varietals with proven Certificates of Analysis (COAs) that are greater than 15% total CBD and less than 0.3% total THC in field trials should be identified. Also, varietals containing greater percentages of alternative cannabinoids create products with higher values for the marketplace and are in greater demand.

Genetics can be weighed against each other via proven COAs. Multiple years of proven COAs on any given strain or varietal is what separates experienced breeders from those looking to join in on the current cannabis rush.

The most important aspect to consider when selecting genetics is the ratio given on the COA, specifically the ratio of CBD:THC. Ideally, this number is greater than 30:1. If you divide 10, which is considered a minimum field percentage of CBD to produce commercial grade product, by 0.39, which is the greatest legal percentage of THC allowed in the state of Wisconsin, you will see that ratio is 25.6:1.

The greater the ratio, the more time the crop can be given in-season to add CBD without adding too much THC. Any ratio greater than 30:1 is considered superior genetics in the current cannabis space (as of the writing of this book). Breeders are continually selecting parentage genetics to increase this ratio for future crops.

The same basic principles apply to alternative cannabinoids, which can all be weighed in this same format. The starting number of 10% as a commercial standard may differ when considering alternative cannabinoid content. Also, some modern varietals targeting alternative cannabinoids contain untraceable amounts of THC.

Managing the THC testing in season is important to ensure that the crop passes state laws surrounding hemp cultivation. Any varietal with a COA at or less than the 25.6:1 ratio must be managed to keep the ratio balanced. The majority of current CBD cannabis genetics have similar CBD- and THC-production curves. In most cases, the CBD concentration approaches 10% at or near the time when the THC concentration approaches 0.39%. When using genetics that are close to the minimum commercial ratio, growers must be sure to have state testing (done by the Department of Agriculture, Trade and Consumer Protection in Wisconsin) conducted prior to the crop passing the 10% CBD mark.

Knowing the market that the grown biomass is intended for may directly affect selection criteria. Within the industry, certain strains have been earmarked for alleviating pain, aiding in sleep, curbing anxiety, etc. This may carry weight when identifying seeds and/or clones for selection. The specific terpene profile that comes from a strain may help with selection as well. The smokeable flower market is also dictated by those strains that produce high-quality buds that smell and taste according to user standards. Identifying strains that are in demand starts with understanding the market where the flower is being sold. Calling or asking store owners or wholesalers which product moves the most volume or has the greatest demand will go a long way toward staying relevant with the current consumer trends.

Feminization SOPs should be researched and obtained from the breeder to ensure proper purity. Reputable sources include: Oregon CBD, Tesoro Genetics, Jupiter Seed, Boring Hemp Company,   Blue Forest Farms, Phytonics, Phylos Biosciences, High Grade Hemp, KLR Farms, and many others.

From personal experiences, some seed companies struggle with purity and feminization. Those that conduct business indoors and utilize a "mother crop," which is propagated from to produce identical females, are likely to have high purity. Also, a certain percentage of those propagated females should be "hermed out," or forced to produce hermaphroditic pollen sacs. This is accomplished with the application of silver nitrate and/or silver thiosulfate to the propagated females at specific times of their vegetative growth. Using pollen from those propagated hermaphroditic females to pollinate other propagated females will produce feminized seed of high-percentage purity if a select number (ideally one) of the mother stock is the originating source. Increasing the number of mother plants with slightly differing characteristics (think sisters) increases the phenotypic differences in corresponding feminized field grows for commercial production.

Feminized seed from reputable breeders carries with it third-party testing results that prove the feminization to be greater than 98%, with most at more than 99.8%. Those percentages equate to approximately one male per acre or one male per 10 acres. This low rate likely is more economic vs. the added cost of clones.

Clones

Cloning of cannabis plants sounds much more sophisticated and technologically advanced than it is. In simple terms, it is the propagation of one single plant into many others. A cutting of a live "mother plant" is taken. That cutting is planted into its own tray within a greenhouse setting, and a rooting hormone is added, so that the specific cutting establishes its own set of roots. Once rooting occurs, the cutting utilizes photosynthesis to feed itself and a new "clone" is the result.

One practical approach to using clones vs. seed starts is that clones are female. There is no need to scout for male removal or have the increased risk associated with having males present within a field to pollinate females. With clones, all of the plants are the same, and all are female.

However, clones tend to be less hardy than traditional seed starts when it comes to general disease and pest resistance. This is because the root system of clones is weaker than that of seed starts. The weaker root system is the result of the cutting/rooting hormone approach vs. the natural seed germination/root growth approach.

CHAPTER 4

Field Preparation and Selection

Field preparation

When growing cannabis, the focus during field preparation should be to maximize seed-to-soil contact. This is similar to other small-seed plants. Prior to planting, selected fields should be aggressively tilled to create soil that has a fine particle size. To manage this fine particle size, try using an adapted transplanter with rollers on the front of the row unit. They are better-suited to firm up the seedbed prior to transplantation of the plug. Rolling post-planting may increase germination rates of direct-seeded fields.

Use pre-plant fertilizers that maximize early-season growth and that will not leach out throughout the growing season. Specifically, potassium (K) fertilizers can be applied pre-plant with success. Cannabis utilizes K throughout the growing season for lignin tissue growth, which mobilizes other nutrients throughout the plant, and for flower and seed formation. If not carefully monitored, potassium can easily become the limiting nutrient. The challenge becomes delivering Organic Materials Review Institute (OMRI)-approved K with economical timing for plant needs.

Chicken litter can be used as a source of nitrogen (N) for organic farming.

CAUTION: Chicken litter contains large amounts of calcium (Ca 2+). Although the plant needs calcium, application of chicken litter at rates needed for N supplementation may result in over-application of Ca. Calcium (Ca 2+) is a positively charged ion in solution and binds to the same receptors to cross cell membranes as K+. If over-application of Ca occurs, the plant will soon show K+ nutritional deficiencies. Increased application of K+ may help, but likely the plant will suffer from K+ deficiency for the entire growing season. Use of chicken litter from broilers vs. laying hens will likely avoid this result.

Certified organic nutrient sources

Nitrogen can be applied pre-plant to promote early-season vegetative growth. Using N that is readily available to the plant, which means it doesn't have to pass through the nitrogen cycle, is most advantageous. Only a portion of the total N requirement needs to be available early in vegetative growth because the plants are small and don't yet require the full amount.

"Spoon feeding" N throughout the growing season should be adopted when possible. A reliable OMRI-approved source of N is Chilean Nitrate (15-0-2). Application of 100#/A of Chilean Nitrate provides 15 units of N and two units of K for early-season plant growth. This nitrogen is in nitrate form, so it's readily available to the plant once in solution.

Certain micronutrients should be applied either pre-plant or early post-emergence. Sulfur, zinc, manganese, boron, and calcium are utilized within the plant. When transplanting, all of these micronutrients may be applied in a liquid drench form.

If direct-planting, applications could occur in a pre-plant side-dress form (liquid) or an early post-emergence form (liquid foliar). If applying micro-nutrients in drench, pre-plant side-dress, or post-emergence foliar forms, tank mixes including sugars (terrafed) and fish emulsions (liquid N source) should be considered. Products containing kelp (N source) are also safe to use on in-fantile cannabis plants. Other micronutrients of concern include molybdenum and calcium silicates.

Early-season drench or pre-plant side-dress applications are great timing events for applications of other critical plant defenses. Serenade Opti (bacil-lus subtilis) is a live bacterium that can be soil-applied for plant uptake. It can provide systemic protection for certain diseases including: botrytis (bud rot), sclerotinia (white mold), xanthomonas, and erwinia. Two of those four diseases, botrytis and sclerotinia, are economically relevant pests to outdoor cannabis production.

Another tank-additive that should be considered at this early application event is that of beauveria bassiana. This is a beneficial fungi that becomes sys-temic within the plant. It creates natural defense mechanisms for the plant when the plant encounters direct insect feeding. A two-pass application pro-gram of soil-applied beauveria bassiana is recommended to create season-long insect resistance, which includes resistance to the eurasian hemp borer.

Seedbed preparation

Because hemp seeds are small and oval-shaped (about one-eighth of an inch in diameter and three-sixteenths of an inch in length), an ideal seedbed around the seed needs to be firm, granulated, and moist (but not so moist as to be sticky). When hemp is direct-seeded with a drill, seeds should be placed at a depth of one-quarter of an inch in loamy soils. Clay soils should be planted at one-quarter of an inch or shallower, whereas sandy soils should be planted from one-quarter to one-half inch deep. Muck and peat soils should be planted at the one-quarter inch depth.

Once hemp seeds start to first absorb moisture from the soil, the soil must not dry out prior to emergence. If it does dry out surrounding the germinating seed, the seedling may perish before it ever reaches the soil surface. That is why it is important to plant into moist soil, where the seed is wrapped in moisture to the surface. There should be no air pockets surrounding the seed, which can happen when planting into cloddy seedbeds.

Soil should be tilled to a depth of four to five inches, with primary tillage happening prior to planting to loosen it and kill weeds. Implements for tilling the soil to that depth include disks, field cultivators, and chisel plows set to a shallow depth. A specialized disk known as an off-set disk will have an advantage when one is tilling a grassy sod for the first time. Such a disk is aggressive at tearing the sod clumps apart due to the steep angle of the disk gangs as well as the deep notches in the blades. If the soil is on the moist side at the time of primary tillage, the soil should be allowed to set for a day or two to air dry. That process may also help to kill some of the large weeds that were present. Primary tillage will also serve to incorporate fertilizers such as poultry manure, aglime, and other soil conditioners.

Secondary tillage should then be performed with a goal of re-firming, leveling, and granulating soil in preparation for seeding. Tillage tools that can be used for secondary tillage include shallow field cultivators with trailing rolling baskets, shallow disking with a trailing spiked harrow, cultimulchers, soil rollers, and vertical tillage tools.

If the soil moisture is on the dry side at the time of primary tillage (and there are only a few small two- to four-inch weeds present), combination tools such as disks with a trailing spiked harrow or field cultivators with a trailing rolling basket or spiked harrow can be used in a one-pass approach. If the soil is nearly free of any spring weed growth, vertical tillage tools (which do not uproot weeds well) can be used in a one-pass approach to seedbed preparation.

An additional pass with the relatively new three- to four-foot diameter soil roller can effectively refirm and level the soil surface after previous tillage. Ideal soil surface firmness can be described as soil that depresses only about one-half to one inch in the heel area of someone walking over the tilled soil.

Crop rotations

Crops have been rotated from one species to another in subsequent growing seasons for decades as a tool for increased crop safety, efficiency, and yield. The reasoning is simple; disease pressure, insect pressure, nutrient deficiencies along with many other negative carryover effects, plague crops that are continually grown in sequence on the same acre year after year. To combat this, crops are rotated from one species to another to avoid these practical negative effects.

Feminized hemp that is intended for the production of cannabinoids should be grown on acreage that has no history of recent residual herbicide or fertilizer usage. These additives, if available in the soil, will most likely be taken up by the hemp crop and be deposited into its biomass. Once the crop is further processed into oil, the contaminants risk adverse effects to human health. This will be tested prior to oil production, and any material containing harmful contaminants will be terminated.

Growers cultivating hemp crops should use this practical knowledge to focus specifically on the correct crops to rotate with. The crop that precedes hemp can be especially important in providing specific nutrients that will be available to the growing hemp crop. We believe that the ideal crop to rotate toward hemp is established alfalfa. Alfalfa is a legume species. These species fixate nitrogen in their root structure utilizing specific bacteria that create a mutually beneficial relationship for the alfalfa and bacteria.

Terminating old established alfalfa stands will provide immediate nitrogen that is readily available to the growing hemp crop. Growing hemp after other crops is possible, but other previous legume crops are preferred. The previous legume crop could supply as much as 60 to 120 units of nitrogen to the coming hemp crop based on quality of alfalfa stand prior to termination. This is the entire requirement for the hemp crop.

Basically if growing after a high-quality stand of alfalfa, it is possible that you will not need to further supplement additional nitrogen to the following hemp crop. This is unlikely; however, it is more likely that you will only have to apply a small amount of the total requirement of nitrogen for supplementation.

Growing hemp after corn or sod (grass) will require near the maximum amount of supplemental nitrogen needed to grow the crop. This is because the previous corn crop utilized all of the available nitrogen in the soil, and there is little available for the growing hemp crop. Hemp can also follow winter wheat in the rotation, but the effects are the same because wheat will likely have consumed the majority of the available nitrogen in the soil the previous year.

Hemp grown after soybeans or potatoes carries with it an extra risk for disease infection. Diseases such as septoria leaf spot infect potatoes annually and are likely remaining in the soil the following year.

Cannabis is extremely susceptible to septoria leaf spot, and a serious economically relevant infection could be the result. Growing after soybeans carries with it increased risk for white mold (schlerotinia) infection. White mold spores will definitely be present following soybean crops that have been infected with the disease. The following year's hemp crop is at extreme risk for schlerotinia infection that could easily result in economical damage.

CHAPTER 5

Soil and Seeding

Understand how to direct seed

There are both positives and negatives to keep in mind when considering if direct seeding is the planting approach you'd like to use for flower production. The positives include: less cost and labor for planting, no cost for plug and clone propagation, and reduced planting duration. The negatives include: waiting for in-season soil temperatures to increase (likely June planting), which decreases the growing season, as well as germination loss and the requirement of specialized planting equipment.

Direct-seeded cannabis for feminized flower production should be seeded no deeper than one-fourth to one-half of an inch. The three main variables that can be managed to maximize the germination rate of the seed are establishing good seed-to-soil contact, soil temperature, and soil moisture.

The population selection for each strain differs based on varietal (indica vs. sativa), row spacing, and expected germination rate. Specific planters that are able to singulate small seeds are absolutely paramount to achieving success when direct seeding feminized cannabis. The current industry-standard cost of seed, which ranges from five cents to $1 per seed, requires consideration of those factors when establishing a live crop from direct seeding. If utilizing direct seeding for final stand establishment, it is of even greater importance to acquire seed at a valued price point to ensure an adequate final stand after loss. Utilizing cheaper seeds of equal genetics will help to offset the difference in loss vs. transplanting.

The use of mechanical rollers is one adaptation to increase the seed-to-soil contact of direct-seeded cannabis. Rollers that "move" soil, or have ridges and valleys on the rollers, should be avoided. Those roller types will likely bury seed deeper than is recommended. Deeply planted cannabis seeds won't have adequate oxygen supplies to germinate, so even if soil moisture and temperature are sufficient, the seeds will not sprout. Flat rollers that are similar to that of lawn rollers are ideal. These rollers are hydraulically raised and lowered, filled with water to add weight, and pulled with a tractor through the field. The weight of the unit packs down the soil by eliminating air spaces between soil particles. These air spaces become present when soil is tilled, and removal of them will likely increase seed germination-rates.

Gibberellic acid, which promotes growth, can be applied to cannabis seeds as a treatment to increase germination rate. We have not tested this hypothesis in detail; we have only researched the limited (at best) data surrounding "old" cannabis seeds intended for high THC grows. There is some scientific data on the use of gibberellic acid as a seed germination promoter in other plant species, and the usage rates and application methods would be similar based on seed size. Again, we have no direct experience, and the data is murky. Some data tends to show that gibberellic acid may decrease the amount of CBD produced in cannabis flowers and increase the amount of THC in cannabis flowers. If that is truly the case, then the increased germination rate would not be enough to offset the increased likelihood for crop failure due to THC-limitation testing.

If a cover crop is pre-planted on the field for weed suppression, great care must be taken to ensure that rows are cleaned of residue and debris when planting. In order to ensure you can properly manage the cover crop, consider using winter rye planted in the fall to provide a crop to suppress weed growth in the spring. This cover crop tends to be terminated in three main ways: natural winter termination, tillage, or roller-crimping. Any and all of the previously mentioned techniques can adequately terminate an established cover crop.

What happens next is especially important for the following hemp crop. Large amounts of residue over the top of the row can smother, shade, or choke out the infantile hemp seedlings as they emerge. Row cleaners can be utilized to clear the row of unwanted debris prior to planting. This is recommended due to the high-cost nature of hemp seeds.

Seeding rates for feminized grows vary based on genetic selection. Sativas should be seeded at 2,500 to 2,750 seeds per acre to ensure a final stand near 2,000 plants per acre. This is factoring in an 80% germination success rate. That number is highly variable based on genetics, soil temperature, planting depth, and other factors. Indicas should be seeded 3,500 to 3,750 seeds per acre to achieve a 2,500 to 2,750 final stand. If the proven direct-seeded germination rates are greater than 80% these recommended seeding rates can be significantly lowered.

Feral hemp in the state of Wisconsin germinates and emerges mid- to late April. We witnessed and took pictures of the varietal on our own farm emerging the week of April 20, 2019. The average soil temperature for that time in the state is between 50 to 60 degrees Fahrenheit. For this reason, it may be possible to direct seed feminized hemp early in May to maximize the growing season without the added cost of starting the plants in a greenhouse.

This theory has yet to be tested in the state because of the high cost of feminized seed and the risk of poor stand quality resulting from lowered germination rates. We do have partnered growers who have had successful grows from direct seeding in both 2020 and 2021 production years.

Using a greenhouse

When beginning feminized cannabis seeds in a greenhouse setting, there are many management considerations one must consider in order to increase the likelihood for success. Specifically, understanding the overall process from start to finish and working through logistics is critical. The overall aim is to begin with feminized cannabis seeds and end with six- to 10-inch hardened plants that are viable and ready for transplantation to the field.

Equipment and materials

The first factor to consider is the greenhouse itself. Any supplier can adequately size a greenhouse based on the grower's tray size and tray count. We used 72 count trays and have found them to be the industry standard.

Next, be sure that the greenhouse is built properly and is secured properly. It took three of our own laborers one week to build our greenhouse, which is 6,000 square feet. Some greenhouses have roll-up side options; we highly recommend these for the "hardening off" period, which is essentially exposing the plants to wind. Extra lighting and temperature control may be especially critical if a grower is starting plugs early, like we did. Our first seeding was the final week in April.

The lighting can be set on timers to turn on and off, which basically lengthens the daylight in early spring. Temperature regulation is a must to ensure that the germinating seeds are not exposed to excessively cold temperatures. Low temperatures can negatively affect germination rates. We purchased simple red lighting that our greenhouse supplier recommended and liquid propane (LP) burners at each end of the house with a thermostat. Our greenhouse came standard with a simple drop-down watering mechanism.

After securing the greenhouse specifics, move onto the materials required for cannabis seed starts. Those materials are: feminized seed, a medium (either certified organic or not), and vermiculite. The seed is self-explanatory; the medium is what the seed will be grown in. The vermiculite is a mineral that helps to hold soil moisture.

When adequate amounts are procured to cover all of the trays that are needed for the grow, then move on to acquiring the necessary tools to accomplish the task in the greenhouse. Those materials are: a cement mixer, a dibble board for depth, containers to move wetted medium, and vacuum seeders.

Vacuum seeders can be sourced from greenhouse-equipment suppliers, and they should be specially designed for the seed and tray size. We borrowed an old 110-volt cement mixer for a few weeks. The containers are likely poly-bushel baskets with rope handles for ease of use, and the dibble board can be purchased from any greenhouse-equipment supplier.

Seeding trays

Step one: Place 10 to 12 gallons of soil into the cement mixer. Turn it on, and use a hose with a spray nozzle to hydrate the soil. This is a scenario based mostly on "feeling" how moistened the soil is. It should not be drenched, but the seed should not be placed into dry medium. Then pour the soil out into a bushel basket. This will require one worker.

Step two: Add soil to the trays; quickly brush excess soil off the top of the trays to make even. Then pick each tray up off of the table, and drop it eight to 12 inches back onto the table to "pack" the medium into the cells of the tray. This step will require two workers.

Step three: Place the dibble board over the top of the tray, and press down firmly to a depth of three-eighths of an inch. This will create small dimples in the medium, so the seed can be placed into them. The dibble board will need to be periodically cleaned to remove build-up of excess material. This step will require one worker.

Step four: The tray is then seeded with a vacuum seeder. The tray is passed from the dibble-board operators to the vacuum seeders, who singulate 72 seeds to each tray, and add them to the dibbled-media trays. This step will require two workers, one for each of the vacuum seeders. This step will likely be the limiting factor for production. The persons operating the vacuum seeders must take great care to ensure that all cells contain only one seed and not excess seed.

Step five: The tray moves to quality control, where three workers manually push each seed into the medium firmly with a finger to ensure seed-to-soil contact. This quality-control step also ensures that each cell contains one single seed.

Step six: Run small amounts of moist soil through a screen to ensure that there are no clumps remaining within it. Use this "screened" soil to add a thin layer of soil over the top of the seeded trays. This step will require one worker.

Step seven: A thin layer of vermiculite is then placed over the top of the tray. Growers can choose to add a beneficial mite pack to the trays at this time to eliminate thrips pressure if they deem it necessary. The tray is then carted off to its place in the greenhouse.

When the entire area that is intended to be seeded has been completed, then watering must ensue to ensure germination commencement. This watering should continue for approximately 25 minutes to ensure adequate moisture is available for germination. Continue to water two to three times per day at 10 to 15 minutes per session. Ensure that the top of the vermiculite always stays moist during the germination period.

Following these steps with 10 laborers would successfully seed 25,000 to 30,000 plant starts in the greenhouse per day. The protocol can be adjusted as necessary based on the size of the grow.

Remaining greenhouse time

The first 48 hours following seeding are the most critical for successful germination rates and likely uniform growth throughout the year. For us, those that emerged uniformly in trays in the greenhouse continued to follow that path in the field and created even full stands. Ensuring adequate water availability and temperatures that are consistently in the 75- to 85-degree range will create an environment conducive to high-germination rates. Usage of "germination chambers," or specially designed areas to hold the trays in during germination, will also likely increase the overall germination rate.

After the germination period ends, the plants are simply growing and strengthening. The grower is through the most critical portion and now needs to shift focus to setting the plugs up for the transplantation event. Ideal plants for transplantation are six to 10 inches in height.

Short plants tend to always be behind and struggle to take root. Tall plants create problems for the transplanter itself and could be injured during the process. Those tall plants also carry with them an increased risk of becoming root bound, which means a large root mass has expanded throughout its limited space and has nowhere to grow.

Total time in the greenhouse to develop the plugs will likely be three to four weeks. During the first week, the plugs will germinate via sending out a radicle, or first root growth. This radicle will bury its way into the medium and establish itself as an anchor point. From there, the plant will push its own seed up and out of the medium and into the sunlight. This takes energy from a little seedling, and to do so requires both adequate moisture and temperature. The growing seedling will push its seed coat up and out into the free world. From there, the two cotyledon leaves spread apart and split the seed coat, which will likely fall off at this point. These first leaves contain the energy the seedling is using for growth prior to photosynthesis development.

Cotyledon leaves are important to the overall health of growing/establishing plugs. The axial growing point is located directly between the two cotyledon leaves. It is the point at which the next set of leaves will emerge from, or the growing point of the plant. As new leaf sets emerge and develop, the plant slowly begins to create and provide energy for itself via photosynthesis. Prior to this the two cotyledon leaves will slowly be consumed as an energy source. This is similar with many other plant species.

The cotyledon leaves will eventually turn brown or grey, become brittle, and fall off. We noticed during actual transplantation that after three weeks of growth in the greenhouse, most of the cotyledon leaves were already brown. We then buried the transplanted plugs right up to the collar of the cotyledon leaves. This extra depth of planting helped to keep the new transplants standing upright after transplantation.

During weeks two through four, the plant is growing. Again, a six- to 10-inch plug is desired for transplantation. This plug has a "hardened" stem. In the early weeks of growth, that stem is fragile and thin. To thicken it, expose the seedlings to prevailing winds during the day. We accomplished this by simply rolling up the sides of our greenhouse and allowing the prevailing winds to cross the space. Another option would be to carry all of the trays outside during a specific time frame to harden the plants off. This wind exposure will force the plants to add extra lignin to their stems for added strength. We noticed that it also slowed the overall vertical growth of the plants that were "stretching their legs." The hardening-off period can be executed during the final week of greenhouse growth, just prior to transplantation.

Light exposure can be artificially supplemented during this period, so the plants are only exposed to a short night period. This can be as short as six to eight hours of darkness in one 24-hour day.

The watering schedule can and may change based on the overall weather outside. Wet days add humidity to the greenhouse. Long dry periods will need extra attention. Expect to water the plants two to three times per day at 15 minutes per time. This watering should be just enough for the medium to stay moist, but there won't be continuously leaking water out of the holes in the bottom of the trays.

Having ventilation in the greenhouse will be helpful during times of excessive humidity. We found that after rains or on certain mornings that the greenhouse would fill with fog. This needs to be eliminated as quickly as possible because long duration exposure could lead to early-season disease pressure.

Soil selection
Soil types
Cannabis is a diverse and adaptive plant. It can be grown across many different soil types with success, but understanding which types increase the likelihood for greater yields can be extremely economical decisions. Those soils that have adequate natural drainage, have high water-holding capacity, and high levels of phosphorus and potassium should be targeted for an economic yield. Also, growing hemp following alfalfa has the added benefit of carryover nitrogen availability.

Soil consists of varying parts of silt, sand, and clay. Whichever of these particles dominate the soil matrix helps to identify which soil type predominates within the field. Soil maps specific for each state or county are free and publicly available.

Ideal soil types for plant cultivation are silt loams. Silt-loam soils contain 70% silt and clay particles and 20% sand. The remaining material is both organic matter and inorganic matter (rock) particulate. A grower's personal understanding of what soil type he/she is growing on will help with decision-making plans, along with any adjustments that follow.

Ideal soil pH for growing cannabis ranges from 6 to 7. pH outside of this range should be adjusted as necessary for optimal production. Some soil types that have been exposed to aggressive row-crop production and high-salt fertilizers become acidic throughout time. To combat this issue and increase soil pH back into the optimal range for production, farmers should apply lime on a basis of one ton per acre, as necessary.

Lime comes in varying forms and concentrations, and it's usually priced accordingly. Keep in mind it takes time for surface-applied lime to break down and interact with soil particles to alter soil pH. This will not happen overnight. Liming can be accomplished in early spring or late fall after the season's crops have been removed.

Organic matter (OM) is a measure of all decaying and stable plant matter within soils. Most soils in the state contain approximately 2% to 3% OM. Those soils that have high concentrations of organic matter have an increased likelihood of supplying crops with the nutrients they need during times of stress. Soils with more organic matter have greater nutrient availability, and therefore, they mobilize both macro- and micronutrients more efficiently throughout the soil matrix creating more plant uptake.

As previously stated, cannabis can be cultivated on a quite a range of soil types. The plants will need adequate nutrients and moisture to grow and thrive. We have a first-hand understanding of many of the soil types where cannabis production occurs. In 2019, we grew not only our feminized production, but also we grew half of our dual-purpose grow on sands with overhead irrigation. They were pivot irrigation fields with mainly sand-specific soil types.

Although we were successful in our grows, we quickly realized that soils with more silt or clay particles and less sand would be better-suited for hemp production. Sandy soils have low organic-matter percentages (usually less than

1%) and are therefore unable to "hold" soil nutrients long-term to be available for continued crop growth.

Alternatively, we grew 35 acres in heavy clay soils that were non-irrigated. These soils have the ability to hold soil nutrients, but also they may become "locked" if moisture is not adequate. Clay soils, when they become dry, tend to bind tightly to nutrient particles, which makes them unavailable to the growing crop.

Fortunately, in 2019 we received adequate precipitation throughout the growing season. Looking forward, we recommend that feminized production acres have the ability to be irrigated to ensure success. Dual-purpose acres, which are of much-lesser value, will not necessarily need to have irrigation for successful production.

Soil testing

Soil testing is a fairly common practice amid modern agricultural practices. Soil should be tested for the standard set of macronutrients that hemp requires, including: P, K, and pH. Soil should also be tested for micronutrients: S, B, Mn, and Zn.

Organic matter is especially important because we have found that, similar to other crop species, cannabis excels when planted on soils with large amounts of organic matter. Soils with more than 3% organic matter are ideal.

Also, soils can be tested for CEC for a better understanding of nutrient availability. These standard testing criteria should all be applied to hemp and used to reference against nutrient requirements for overall fertilizer-application plans.

Outside of the usual testing for row crops, soil intended for cannabinoid production should be tested for other contaminants. Specifically, the soil should be tested for four heavy metals: lead, cadmium, mercury, and arsenic. These tests are also available as water tests. Test the water that's to be used for irrigation throughout the season for those heavy metals.

Testing the soil for heavy metals can be easily missed with all of the management considerations that need to be addressed when cultivating cannabis. This step should not be taken lightly. Tests with low or not-present results are ideal. It is not yet completely known what levels of heavy metals will result in biomass or oil tests that are greater than the threshold for safe human consumption. We suggest erring on the side of caution here.

Residual herbicide testing is available from some soil-testing laboratories. The tests can be expensive, and usually the grower will have to list the contaminants to specifically test for. There are about 400 pesticides that hemp biomass and cannabinoid oil will be screened against to ensure human safety-consumption requirements are met. For this reason, it is ideal to identify acreage that is low risk with a proven history of limited- to no-pesticide usage. USDA-certified organic acreage is ideal acreage to target to ensure production standards are met.

If a grower has no choice but to test fields that have a history of pesticide application, we recommend obtaining a complete understanding of the specific pesticides that have been applied. The reason for this is certain pesticides stay within the soil longer than others.

Residual herbicides that bind to soil particles are the most critical to understand. Herbicides with trade names such as Lumax, Acuron, Callisto, Realm Q, and more are likely to be present in the soil for 12 months or longer post-application. It's a critical step to research into the specific history of pesticide usage on specific fields that are intended for feminized hemp production.

CHAPTER 6

Transplanting

Planning

Transplanting hemp plugs can be a critical portion of success when growing hemp for cannabinoids. Once growers determine that transplantation is their preferred method, planning must begin. Schedule the seed starts or clones based on a "target" transplant date. May 20 is a recommended target date to consider beginning transplantation of hemp in Wisconsin.

Hemp plants can tolerate cold temperatures, but newly transplanted hemp that is exposed to a late or difficult frost is likely to either suffer from transplant shock or spend a certain amount of time in "recovery" mode vs. rooting or taking root. Changing the actual transplant date based on current expected weather patterns is advised to avoid crop injury.

Seed starts will require anywhere from three to five weeks of growth in the greenhouse before an established 6- to 10-inch plug with adequate root mass is developed. Be sure the seedlings have been "hardened off," or exposed to periods of prevailing winds to stiffen their stems, and ensure they will not topple over once transplanted. Also, we found that if the plants have been held back in their watering in the greenhouse to help harden them, a good soaking of the trays just prior to transplantation will help to hold together their root masses after pulled from the tray.

Another option to consider would be plug "poppers," which is when the grower manually forces a metal pin through the hole in the bottom of the tray to help the plugs release from the tray itself. This should be used as a last resort because it may also lead to root damage and therefore, some transplant shock. We used plug poppers on our hemp plugs with little transplant shock, but it is an added step and increases labor cost and timing throughout. If starting the plants yourself, plan backward from the intended plant date to time the seeding of the trays in the greenhouse. The same logic can be used if a grower is paying another entity to begin the plugs.

Planning must take into consideration the time required for transplantation itself. A labor force is required to move the plants themselves, from wherever they were started, to the field. Labor is necessary to drive the transplanting tractor, along with one person per row unit. We've learned that a person walking behind the rig can be avoided by properly adjusting the transplanter's settings.

If plugs are being improperly placed, have a trailing person "fix" or adjust planted plugs as necessary. If one acre can be planted in one day and five acres are intended to be planted, then seeding the trays one acre at a time for five consecutive days in the greenhouse is advised. We were extremely cautious in our approach to this final piece, and we wished we had seeded our entire 100,000-seedling greenhouse in one week's time because our four-row transplanter would plant 10 acres per day. Instead, we seeded one-third of the capacity during three weeks' time. We rotated new seeded trays in as sets were transplanted to the field, and we had 240,000 seed starts total. It then took six weeks to finish transplanting when it easily could have been accomplished in three.

Timing and planning is critical to grow large productive hemp plants for flower production. The early-planted varietals were easily the highest yielding of our entire crop. This is similar to other crops grown in the state due to the limited growing degree day units (GDUs) available in the state. For this reason, planting date is critical in ensuring high-yielding productive plants. This is true whether a grower is transplanting or direct seeding.

Ground temperature as well as air temperature for 48 hours post-planting is extremely critical when direct seeding. The first week in June is the earliest date for direct seeding feminized hemp in the state due to those temperature requirements.

Transporting plugs

When transporting live hemp seedlings or clones from the greenhouse to the field precautions must be taken to ensure no detrimental damage is done to the plants. Wind exposure during transport is the greatest risk for plants, along with unintended adverse weather conditions.

Transport can be accomplished in a covered truck or trailer with shelving for the trays to maximize space within. Other variations of this basic concept exist, but transport using open-topped vessels will cause problems if high-wind or hail events occur during transport.

Plants that are bent over during transport become highly problematic when individuals are grabbing plants from trays in the transplanter itself, and they tend to cling to the cup on the carousel of the transplanter itself. This can lead to skips or plants with cut tops post-transplantation.

Ideally, plants would be taken out of their greenhouse environment the day of planting and planted that day. This is not always practical due to the distances of transportation, so batches may have to be moved with a certain portion held overnight close to the field for the following day. Practical approaches and common sense go a long way during this procedure, and this can be implemented successfully a number of ways.

Planting plugs

Always read through the entire owner's manual for the transplanter prior to attempting transplantation. Some concepts are self-explanatory, and some are not. Pulling plugs out of trays on the row unit itself is quite self-explanatory in its nature. Pull the plant from the tray using the strong "hardened" portion of the lower part of the stem, just above the line to the medium, or potting soil. If variability exists within a set of trays based on plant height (fairly common) it is best to divide the trays based on row units.

This is a simple concept.

If it is a two-row planter, adjust one row unit to plant the taller trays and one row unit to plant the shorter trays. If more row units exist, more variability of trays can be accommodated. Modern transplanters have many adjustments that can be made based on variability of the plants. Short plants should be planted shallowly, so a thin layer of soil pressed by the gauge wheels covers the medium. The "pack" wheels will then press the open slot created by the plow on the front of the row unit closed to firm up the plug in its place. If plants are tipped either forward or backward, and not upright, refer to the transplanter's user's manual for adjustments.

Transplanters equipped with a drench are ideal for transplanting feminized hemp. The tanks for the drench can be mounted as "saddle" tanks on the tractor itself. Although, some large transplanters have their own tanks. The liquid in the drench is mostly water, and the amount provided per plug should be available in the user's manual.

Consider additives of sugar (TerraFed at two gallons per acre), beneficial bacteria (Serenade Opti at labeled rate and beauveria bassiana at labeled rate), liquid zinc sulfate (one gallon per acre), liquid manganese sulfate (one gallon per acre), and liquid boron (one quart per acre). The sugar and micronutrients help the plants take root and reduce the likelihood for transplant shock. The beneficial bacteria are taken up by the roots or bind to root hairs to help create systemic pathways in the plants to fight off pest pressures. The Serenade Opti will help with foliar disease pressure (Septoria leaf spot, etc.), and the beauveria bassiana will help the plants to ward off insect pressure including eurasian hemp borer. If using dripline irrigation most of these products can be applied shortly after transplantation with the same effect.

NOTE: Population of transplanted hemp should be based on varietal type and should follow the advice listed in the genetics portion of this handbook.

Many different aspects must be taken into account when deciding to use dripline and/or plastic coverings for controlling weeds in the beds of growing hemp. The first aspect to consider is general weed control.

NOTE: Refer to the weed-control portion of this handbook for the different aspects of weed control available. Plastic is extremely effective at keeping rows clean, but there is the added costs of the plastic, labor, and hassle associated with laying the plastic and its removal after the growing season. Plastic and driplines are effective tools for feminized hemp grows ranging from one to 10 acres in size. Land greater than 10 acres requires an increased skill set and greater management to successfully accomplish both of these practices.

CHAPTER 7

Preplant Fertilizers

Preplant fertilizer usage

Applying preplant fertilizer to fields intended for cannabis production is highly beneficial for those plants to receive adequate nutrients during the growing season. Specifically, plants need the macronutrients nitrogen and potassium and the micronutrients sulfur, boron, and zinc. Whether the plants will be organically grown or USDA-certified organic, specific fertilizer blends are readily available from local cooperatives or fertilizer dealers.

The plant consumes large amounts of nitrogen upfront and large amounts of potassium later in the growing season. Also, micronutrient availability is essential to avoid transplant shock if that's the intended planting type. Specific micronutrients such as sulfur, boron, zinc, and manganese are critically important both early and late in the growing season — early as the plant takes root and late as the plant flowers. Flower development can be adversely affected if the plants are limited in potassium or micronutrients.

Organically Grown - conventional

The main advantage to "organically grown" feminized hemp for cannabinoid production vs. USDA certified organic is that the fertilizer cost is less. Urea, ammonium sulfate, chelated boron, chelated zinc, and potash are common fertilizers for other row crops. They are readily available at low price points. Custom-blended preplant fertilizers can be created specific to the acre based on soil testing.

A basic starting point for a preplant fertilizer for "growing organically" is as follows:

- 100 pounds per acre K-Mag 0-0-22-22S-11Mg
- 200#/Acre Potash (K2O) 0-0-60

- 100#/Acre Urea   46-0-0
- 1#/A Zinc (actual)
- 2#/A Boron (actual)

This preplant fertilizer blend provides most of the plants' basic nutrient needs during the early portion of vegetation growth.

100#/Acre K-Mag   0-0-22-22S-11Mg

The 100#/Acre K-Mag provides 22 units of potassium to the crop and 22 units of sulphur. It also applies 11 units of magnesium. This fertilizer is not volatile, which means it will not bond to atmospheric molecules and be lost in availability to the crop when applied to the soil surface.

200#/Acre Potash (K2O)   0-0-60

The 200#/Acre Potash provides 120 units of potassium or K2O. This potassium is slowly released throughout the growing season. Potassium binds tightly to soil particles and can be "held" within the soil itself. Soil-testing results for potassium are a combination of potassium that is readily available for crop needs and that which is bound to soil particles.

If it is determined that sufficient quantities of potassium are available to the hemp crop from a soil test prior to application of preplant, this fertilizer may not be necessary. Conversely, if the soil testing shows that soil K2O levels are below-average, the amount of potassium that should be applied should increase.

100#/Acre Urea   46-0-0

The coming hemp crop receives 46 available units of nitrogen from the 100#/Acre Urea. This nutrient can easily become the most-limiting in terms of early-season requirements for growth. Urea is volatile in its nature and will bind to atmospheric particles. If days of dry hot weather persist after application, a portion of the fertilizer likely will be lost. Applying it prior to a rainfall event limits this loss significantly.

Urea, along with many other nitrogen fertilizers, can "leach" out of the soil during excessive rainfall events. This nitrogen binds easily to water, so it's readily taken up by the plants during and after rainfall events when the ions are in solution. It's also the reason why it can "leach" out of saturated soils.

1#/A Zinc (actual) and 2#/A Boron (actual)

The 1#/Acre of actual zinc and 2# boron are added to the preplant blend to ensure those micronutrients are available to the hemp crop throughout the season. This application will likely provide sufficient amounts.

Also, the 24 units of sulfur in the AMS should provide sufficient levels of sulfur for the entire growing season. If this basic blend was applied preplant in its entirety to the coming hemp crop, the only other nutrients that would need attention are phosphorus and manganese.

Phosphorus

Total phosphorus (P) needs of feminized hemp are relatively low, 30 to 60 units, considering the other nutrient requirements. Due to these low phosphorus requirements, most soil types contain sufficient levels of phosphorus for crop needs. Application of livestock manure, even at low rates, will provide sufficient levels of phosphorus for the hemp crop. Phosphorus is highly water soluble, so following Wisconsin's 590 Nutrient Management requirements based on field data is essential to ensure environmentally safe applications of manure.

If considering fertilizer, a small amount of di-ammonium phosphate (DAP) or monoammonium phosphate (MAP) would suffice. The hemp crop receives 18 units of nitrogen and 46 units of phosphorus from 100#/Acre of DAP (18-46-0). Nitrogen fertilizer rates may need to be adjusted accordingly.

Manganese

Manganese is the final nutrient that would need to be addressed early on in the vegetative phase. Manganese can be safely applied as a foliar feeding option early after emergence, while plants are six to 10 inches tall, or immediately following transplantation. This micronutrient can help the plant take root in a new soil type and decrease the likelihood for "transplant shock" associated with plug starts. It can also be applied in the transplant drench.

All of the previously mentioned pre-plant fertilizers can be applied via dripline continuously throughout the season to "spoon feed" the plants. When growing with this specific capability, the use of pre-plant fertilizers may not be necessary, and the equivalent units of all elemental nutrients can be slowly added to the root zone for quick plant availability via dripline.

Vegetative Applications

Cannabis maximizes its growth potential and overall yield when it is "spoon fed" nitrogen throughout the growing season. Target 50 units of nitrogen per acre split applied 2-3 times depending on management abilities. This should begin two weeks after the plants "take root" in the field and visibly start to add vegetation.

A pre-bud fertilizer application can be essential to ensure the plants do not become nutrient deficient late in the growing season depending on soil type. This will likely occur in the last weeks of July for photoperiod flowering cannabis varietals. This pre-bud application follows:

- 100#/A AMS   21-0-0-21S
- 50#/A DAP   18-46-0 (If necessary based on tissue test)
- 200#/A Potash   0-0-60

This application would apply 30 total units of nitrogen; 23 units of phosphorus; 120 units of potassium and 21 units of sulfur. This application should be enough to finish out the crop and maximize its flower yield potential.

Developing fertilizer applications from soil and tissue tests

Soil test results are reported as:

- **Soil pH**, the measure of acidity of the soil water, where one-unit differences are a 10-fold change in acidity.
    - pH of 7.0 is neutral, pH of 6.0 is 10 times more acidic than a pH of 7.0; pH of 5.0 is 100 times more acidic than a pH of 7.0

- **Soil organic matter** is expressed as the percent of a soil that is non-mineral in content, which is essentially the OM that is a film or granule around the mineral components of the soil. Those are sand, silt, and clay; OM is considered to be prior plant residue that has been broken down by soil microorganisms to the point of stability (humus). OM is a fragile component of soil that must be continually replaced to remain stable in content. It aids in soil water and nutrient retention, helps soils resist compaction, and provides a haven for soil microscopic life.

- **Levels of** ammonium nitrogen (NH4-N), potassium (K), calcium (Ca), magnesium (Mg), zinc Zn), iron (Fe), manganese (Mn), and copper (Cu) are cations (positively charged ion) expressions. Cations are held in soil solution and on the soil cation exchange sites due to the edges of soil particles (colloids) being negatively charged.
  - Think of how a magnet attracts metal fillings on a table top; these cation soil nutrient levels are expressed in parts per million (ppm) of plant-available nutrients (these are not the total levels of these elements in a soil, just the level that are immediately available for plant uptake).

- **Levels of** plant-available anionic (negatively charged ion) nutrients are also expressed in ppm for nitrate nitrogen (NO3-N), phosphorus as phosphate (HPO4-P), sulfate sulfur (SO4-S), and chloride (Cl). With the exception of phosphate, these anionic nutrients are generally repelled by soil colloids because negative charge repels like negative charge. Therefore, these nutrients tend to reside in the soil solution and are subject to loss whenever the soil water drains down through the profile (leaching) after excessive rains. That loss mechanism is becoming more frequent as climate change induces larger and more frequent rain storms

Ideal soil test levels for hemp grown on loam soils in Wisconsin

Soil Test Parameter	Range in Ideal Test Level
pH	6.2 to 6.8
OM	1.8% to 3.0 %
NH4-N	10 to 15 ppm (early season)
NO3-N	21 to 25 ppm (mid season)
NO3-N	15 to 20 ppm (late season)
SO4-S	15-30 ppm
P	30 to 40 ppm
K	140 to 170 ppm
Ca	600 to 1,000 ppm
Mg	100 to 500 ppm
Zn	3 to 20 ppm
Mn	11 to 20 ppm

The nutrients copper, iron, and chloride are not considered to be limiting in Wisconsin. Those nutrients can be limiting in western states, where soil pH levels can be as high as 8.0 in the presence of sodic (high-sodium content) soils.

The task of turning soil test levels into fertility recommendations is indeed an art and a science all rolled into one. For the sake of simplicity, we will focus on loamy soils. We will take each nutrient one at a time.

- **Soil pH** - If the soil test for pH is less than the ideal range of 6.2 to 6.8 for hemp, aglime will need to be added to the soil to increase pH level. In Wisconsin the following equation is used to calculate the amount of 60-69 neutralizing index (NI) aglime required to increase pH from 6.0 to 6.8.
  - *Example:* 72.7- (7.59 x buffer pH) – 3.78 x water pH
  - Most loamy soils in Wisconsin would require about three tons of 60-69 NI aglime to increase soil pH from 6.0 to 6.8. The amount of aglime required will be calculated and printed on the soil test report.

- **Phosphate (P2O5)** - If the soil test for P is less than the ideal range of 30 to 40 ppm for hemp, then the amount of required P2O5 is as follows:
  - For every one ppm that the soil test is less than 30 ppm, add 18 lbs/A of P2O5 from fertilizer and/or animal manures

- **Potash (K2O)** - If the soil test for K is less than the ideal range of 140 to 170 ppm for hemp, then the amount of required K2O is as follows:
  - For every one ppm that the soil test is less than 140 ppm, add seven lbs/A of K2O from fertilizer and/or animal manures

- **Ca and Mg** - These nutrients are added to loamy soils when dolomitic aglime is used to increase soil pH; therefore, those two nutrients are adjusted whenever aglime is added. That goes hand-in-hand with soil Ca and Mg levels:
  - When soil pH is low, one can expect Ca and Mg to also be low in the soil.
  - Whenever soil pH is adequate, one can expect Ca and Mg to be adequate in the soil.

- **SO4-S** - Sulfur availability in soils is controlled by moisture, temperature, and kind/amount of organic matter and crop residues. A guideline is to have one pound per acre of plant-available SO4-S for every six lbs/acre of plant-available nitrogen. Remember both NH4-N and NO3-N are plant-available nitrogen, so add those two together.
  - Assume one would need 25 lbs/acre of SO4-S to produce a crop of hemp. That implies one would need 150 lbs/acre of plant-available nitrogen to produce that same crop of hemp.
  - To increase the amount of SO4-S in the soil, use ammonium sulfate (AMS) as the sulfur source because AMS also contains some nitrogen as NH4-N. AMS contains 21% nitrogen as NH4-N and 24% sulfur as SO4-S, or 21 pounds of nitrogen and 24 pounds of sulfur in every 100 pounds of AMS. To add enough sulfur for a growing season for hemp, apply 100 lbs/acre of AMS.

- **Zn and Mn** - If the soil tests low for these similar micronutrient cations, add one lb/acre of the nutrient for each hemp growing season. Those nutrients should be applied as part of a side-banded fertilizer at planting. Broadcast applications of Zn and Mn tend to be inefficient in terms of hemp uptake. Both of those nutrients could also be applied as foliar fertilizers as part of planned season application.

CHAPTER 8

Nutrient Management

Macronutrient requirements and certified organic fertilizers

Cannabis consumes nitrogen (N), phosphorus (P), and potassium (K) in similar quantities to other crops such as wheat or corn. Specifically, the plant uses N and K for early vegetative growth. After flowering begins, the plant's nutrient requirements change dramatically, and nutrient availability should correspond.

Late-season feminized cannabis consumes large quantities of K, and that nutrient should be readily available in the later portion of the growing season. Cannabis will consume 80 to 150-plus units of nitrogen, 30 to 60 units of phosphorus, and 120 to 200 units of potassium. The nitrogen- and potassium-consumption rates are on the high end for large plants with large productive flowers.

Nitrogen nutrient management should begin with the previous crop rotation. Crops such as alfalfa contain a legume nitrogen credit that can be especially beneficial for growing cannabis.

When complying with USDA-certified organic standards, fertilizer programs are limited, at best, in providing quantities of nitrogen at economical rates for consumption. Chilean nitrate (15-0-2) is available in a dry-granular form that can easily be applied either broadcasted or top-dressed. Nitrogen applications should be applied in a "spoon-fed" approach to have available N throughout the vegetative portion of the growing season when the consumption rates are the greatest.

A useful pre-plant USDA certified organic fertilizer application follows:

- 300#/A Chilean Nitrate   15-0-2   (If no upfront legume nitrogen credit)
- 100#/A K-Mag natural   0-0-22-22S-11Mg   (sulfate of potash-magnesia (langbeinite))
- 200#/A Potassium Sulfate   0-0-50-17S

This single pre-plant application will provide 45 units of nitrogen; 124 units of potassium; 54 units of sulfur; and 11 units of magnesium. We recommend using the lowest total amounts of Chilean Nitrate possible because it adds available salt directly to plants and cannabis is sensitive to high salt content soils. Other USDA certified organic sources of nitrogen can replace the Chilean Nitrate in equal parts.

Applying 100#/A of Chilean nitrate when the plant is 12 to 14 inches tall will "kick start" the vegetative phase. A subsequent application two to three weeks later will continue the "spoon feeding" approach to nitrogen management. Pre-bud applications of N should be considered if necessary based on tissue testing. Cannabis plants with N contents at or near 5% toward the end of the vegetative phase are ideal. A useful pre-bud USDA certified organic application follows:

- 200#/A   Chilean Nitrate   15-0-2
- 200#/A   Potassium Sulfate   0-0-50-17S

"Pushing the plant" with flowering applications of N do not directly cause increased THC within the plant. This misconception exists within the cannabis industry. Plants that "go hot," or become non-compliant, simply have low CBD:THC ratios, with some as low as 1:1. There is little data that supports late-season N applications to increase flower density; although, deficiencies will likely decrease overall yield.

Feather meal is one organic source of nitrogen, with most sources claiming it contains as much as 13% total N required by cannabis plants. However, nitrogen is in an insoluble form in feather meal. If used as a nutritional fertilizer for agricultural crops, it becomes more readily available if mixed with portions of litter. The ideal litter would be from chick or broiler to ensure low levels of calcium that could come from eggshells. Feather meal can be applied in the fall for the next growing season or in the spring, so it's available at the end of the growing season. We would not recommend using feather meal as the sole source of nitrogen due it's low availability to plants.

Chicken litter is another organic source of nitrogen that can be easily handled and applied in pellet form.

CAUTION: Chicken litter from laying hens contains high levels of calcium (Ca), which comes from eggshells. Although calcium is a micronutrient that cannabis requires, chicken litter from hens contains much more calcium than the crop needs. Calcium (Ca 2+) is a positively charged ion in solution, and it binds to some of the same cell receptors as potassium (K+). If over-application of calcium occurs, calcium will likely block the receptors that engage with potassium, and the plants will soon show potassium deficiency. Even if corresponding potassium applications are made, the plants will likely suffer from potassium deficiency throughout the remaining growing season. We have firsthand experience with this mistake, and it's one that should be avoided. Pelletized chicken litter from broilers or chicks is recommended, but it should not be used as the primary source of nitrogen for cannabis crops.

Using manure as hemp fertilizer

During grow season

Manure needs to be applied 120 days before harvest; fall applications prior to the next growing season would work best.

Average Analysis of Liquid Manures (Wisconsin Averages)

Totals per 1,000 gallons of manure	Nitrogen + Phosphate + Potash + Sulfur
Dairy outdoor pit surface-applied (2% dry matter)	4 + 3 + 11 + 0.6
Dairy outdoor pit incorporated within three days (2% dry matter)	6 + 3 + 11 + 0.6
Dairy outdoor pit injected while applied (2% dry matter)	7 + 3 + 11 + 0.6
Dairy indoor pit surface-applied (5% dry matter)	6 + 7 + 19 + 2
Dairy indoor pit incorporated within three days (5% dry matter)	8 + 7 + 19 + 2
Dairy indoor pit injected while applied (5% dry matter)	11 + 8 + 19 + 2
Swine indoor pit surface-applied (5% dry matter)	17 + 14 + 22 + 1.8
Swine indoor pit incorporated within three days (5% dry matter)	22 + 14 + 22 + 1.8
Swine indoor pit injected while applied (5% dry matter)	28 + 14 + 22 + 1.8
Swine outdoor pit surface-applied (5% dry matter)	7 + 6 + 8 + 0.6
Swine outdoor pit incorporated within three days (5% dry matter)	9 + 6 + 8 + 0.6
Swine outdoor pit injected while applied (5% dry matter)	12 + 6 + 8 + 0.6

Example - An application of 8,000 gallons/acre of injected outdoor dairy manure would apply 56 lbs/acre of N (nitrogen) + 24 lbs/acre of P2O5 (phosphate) + 88 lbs/acre of K2O (potash) + five lbs/acre of S (sulfur).

More actual nitrogen per acre will be saved for hemp uptake when the manures are either incorporated into the soil with mechanical tillage or directly injected into the soil at the time of application; that is due to the retention of the ammonia-nitrogen in the manures. Outdoor-stored liquid manures contain less nutrients than indoor-stored.

Other fertilizers

If growing conventional organically grown material but not certified-organic material, then the available nitrogen fertilizer sources are more readily available. Any commercial fertilizer such as urea (46-0-0) can be applied to have 80 to 150+ units of nitrogen available to the plants for the growing season. Once again, spoon-feeding nitrogen to growing cannabis plants is preferred to one single early application. Side-dressing, top-dressing, or injection application practices can be used in the growing season.

UAN liquid solution is a mixture of urea and ammonium nitrate in water. It is possible to use either 28% or 32% UAN liquid solution in season to side-dress growing cannabis plants. Take measures to ensure that any application of banded liquid fertilizers doesn't cause crop injury. Application of high concentrations of nitrogen with high salinity to the root zone can burn root hairs off growing cannabis plants. Apply bands at least eight inches away from plants within the row to ensure plant safety. Similar to other row crops, multiple bands on either side of the row are preferred to a single band on one side of the row.

If utilizing liquid-fertilizer sources to side-dress cannabis in the growing season, the timing of these applications coincide with other applications of beneficial bacteria. Tank mixes including beneficial bacteria (beauveria bassiana), calcium silicates, and liquid potassium should be considered. The high salt content of these fertilizer sources should make them a last-resort nitrogen application strategy.

CAUTION: Cannabis is extremely sensitive to high-salt-content soils or fertilizers, so we consider the use of UAN to be at the grower's own risk.

Nutrient management of phosphorus should be fairly simple when growing cannabis crops. The crop requires only small amounts of phosphorus to reach maximum potential. This requirement is approximately 30 to 60 units of P.

Most soils in Wisconsin, if properly managed and not degraded, contain enough available P for cannabis. Those fields with soil tests at less than 20 ppm (parts per million) should be supplemented. Most livestock manures contain high enough levels of phosphorus to safely apply enough P for the growing season. All applications of manure should comply with state 590 Nutrient Management standards.

Nutrient management of potassium (K), if not correctly managed, can easily become the limiting nutrient regarding cannabis flower yields. Cannabis plants consume potassium throughout the year, and K should be readily available throughout the year to meet those needs. Potassium can be held within the soil. It binds to soil particles, so it does not readily leach after heavy rainfall events.

Potassium can be applied in varying forms. Mined potassium sulfate meets organic-approval standards and can be applied in the fall prior to spring planting or in the spring pre-plant broadcast. If plants require additional K in the growing season based on tissue testing, K can be applied in a dry-granular broadcast form or top/side-dressed next to the row. Liquid organic potassium can be obtained and soil-applied in a side-dress fashion to ensure availability to growing plants. This method is preferred for plants with tissue tests that require additional K. This application is timely for other soil-applied products including beneficial bacteria (beauveria bassiana) and calcium silicates.

Micronutrient requirements

Cannabis requires sulphur (S), calcium (Ca),magnesium (Mg), boron (B), manganese (Mn), and zinc (Zn) as the main micronutrients for growth. Sulphur, zinc, and calcium are utilized during the vegetative phase of growth. Boron, magnesium, sulfur, and manganese are utilized during the reproductive portions of growth. Organic micronutrients can be applied as soil-applied products or foliar-applied products. Liquid forms are readily available and can easily be added to drench applications at transplanting or early foliar applications. They can also be easily applied with dripline applications.

Seed treating is another method of applying micronutrients to cannabis crops. When treating seed, take precaution in understanding use rates and drying procedures in order to correctly establish equal amounts to seed coats.

The sulphur requirement of cannabis can be met with applications of organic potassium sulphate. Foliar-applied sulphur products are available and should be used only if necessary based on in-season tissue testing. Soil-applied products (side-dressed) in season can provide greater nutrient availability to cannabis crops without causing crop injury. Use of soil-applied sulfur applications should only be considered based on tissue-testing results.

Calcium can be applied to cannabis in varying forms. Liquid calcium silicate soil-applied in a side-dress form is preferred in season. This provides available nutrients for uptake in the root zone. Supplementing cannabis crops helps create thick cell-wall formation. This "boosts" plant natural defense mechanisms against predation and disease. Calcium silicate may also protect crops against frost damage. Due to increased cell-wall thickness, the cells of the plants are better-suited to withstand early frosts in the harvest season. These applications may be of greater concern for late-planted cannabis with late-season maturity. Liquid calcium silicates can also be applied via dripline irrigation.

Boron, manganese, and zinc are utilized within the plant at similar stages of growth. Infantile plants (eight to 12 inches) require all three micronutrients to maximize vegetative growth. During budding, flowering, and seed formation, the plant again requires these nutrients to optimize reproductive yields.

Boron, manganese, and zinc are available in liquid organic products and can be foliar-applied. Adjusting pH in the tank is critical when applying foliar nutrients because certain applications carry risk to growing plants. Buffering pH can be adjusted using either emulsified fish or citric acid in the tank. Tank mixes with high pH can cause severe leaf burning and defoliation.

We have found that farmers are inherently independent, creative, and adaptable in their thought processes. For this reason, our advice to all hemp growers is to be creative when considering all applications planned for the hemp crop.

As an example, we applied Chilean nitrate to our entire feminized crop in 2019. To accomplish this we purchased an old used corn planter that was stripped down to the frame with tires, but it still had operational dry fertilizer tanks mounted on the frame. From this, we were able to craft a custom hemp-fertilizer applicator that top-dressed dry fertilizer directly next to the growing hemp plants. All of the components from that old piece of equipment still worked, so we could apply an exact rate of fertilizer on a per-acre basis. Banding of the fertilizer vs. broadcasting made more nutrients available to the root structure of the growing hemp plants. This simple application strategy proved to be extremely cost effective with greater results from the specific fertilizer applied.

Nutrient-deficiency symptoms in hemp

Nutrients are needed in large, medium, or small amounts; those amounts vary within plant parts and stages of growth. There are various ways to tell that a hemp crop is suffering from nutrient deficiency.

Color change in lower leaves is a symptom of nutrient deficiency because nutrients translocate to the top leaves. Colorado State University has a website available showing discolorations directly due to nutrient deficiencies in hemp production.

Nitrogen deficiency

If the plant is deficient in nitrogen, the plant will appear light green with older leaves looking yellow (chlorosis). Yellowing will begin at the leaf tip and extend along the midrib toward the mainstem.

Phosphorus deficiency

If the plant is deficient in phosphorus, it will initially appear dark green with purple casting. The leaves and the plants themselves will be stunted in patches.

Potassium deficiency

If the plant is deficient in potassium, it will have yellow/brown (bronze) discoloration, chlorosis, and scorching along the outer margins of older leaves. This discoloring begins at the leaf tip. It then travels in the same pattern to the middle leaves as plants age and/or continue to be deficient in potassium. If potassium fertilizer is fed to plants, older deficiency symptoms never totally disappear.

Magnesium deficiency

If the plant is deficient in magnesium, older leaves will have yellow discoloration between veins. Veins in the middle leaves will develop reddish-purple colors from the leaf edge inward.

Nitrogen deficiency

If the plant is deficient in nitrogen, it affects the older leaves the most because nickel plays a role in nitrogen transformations (urease enzyme). Terminal growth may cease in young plants, which would produce severe stunting. There may be color change in upper leaves, which means essential nutrients are not translocated to top as terminal buds die.

Calcium deficiency

If the plant is deficient in calcium, the emergence of primary leaves is delayed as terminal bud deteriorates. Leaf tips may stick together in a ladder-like appearance, and new terminal growth appears bubbly on the surface.

The leaves appear yellow/red near the growing point, and growth buds appear as white or light brown dead tissue. On the main stem, the spacing between internodes of leaves are small, giving the overall plant a bushy-top appearance.

There will also be color change in the upper leaves, and essential nutrients will not be translocated to top.

Sulfur deficiency

If the plant is deficient in sulfur, young leaves (including veins) turn pale green, and the rest of the plant turns from pale green to yellow. The plant becomes stunted in growth, and the extreme edge of the middle leaves have red lines.

Zinc deficiency

If the plant is deficient in zinc, there will be chlorosis between the veins on upper leaves, with white bands appearing on both sides of the midribs. Plants may be stunted with shortened internodes.

Iron deficiency

If the plant is deficient in iron, it will show interveinal chlorosis up to the leaf tip on newer leaves, and yellow to almost white leaves eventually appear.

Manganese deficiency

If the plant is deficient in manganese, leaves from the middle to the top will appear yellowish-gray to reddish-gray, while leaf veins remain green.

Copper deficiency

If the plant is deficient in copper, young leaves will appear uniformly pale yellow. They may eventually wither and die without going chlorotic. Buds will not form on the ends of stems.

Chlorine deficiency

If the plant is deficient in chlorine, the upper leaves wilt during dry and hot weather, and then they will turn chlorotic.

Molybdenum deficiency

If the plant is deficient in molybdenum, young leaves wilt and die along the margins. There will be chlorosis of older leaves due to the inability to properly utilize nitrogen.

CHAPTER 9

Foliar Applications to Hemp

Feminized hemp that is intended for cannabinoid production must have OMRI-certified products applied to it during its vegetative growth. This is true of the organically grown and the USDA certified-organic production types. Not only does the product have to be OMRI-approved, but also it must be on the state list of approved products for cannabis.

Cannabis has been illegal in this country for decades, and so it is not found on most (if any) labels of OMRI-approved products. This will likely change quickly, but the manufacturer of the product must first change the label, and then it must be approved via the corresponding state criteria. If hemp is on the label, it is legal to apply in any state where hemp production is also legal. The approved list can be found by contacting your state agency; in Wisconsin that is DATCP.

Start with identifying the application rig to be used in the field. Nozzle spacing and number of nozzles should be based on overall row spacing and number of rows covered by the application equipment. Product usage can be minimized, along with overall cost, if application occurs solely over the rows of hemp. Specific nozzle selection should target the total application volume of 50 gallons per acre early in the vegetative phase of growth. This could be reduced with a narrow fan nozzle. We advise using a fine nozzle that will decrease droplet size and provide good coverage.

Drift concerns associated with large spray droplet size are decreased because only OMRI-approved products can be applied to hemp. These products tend to have little, if any, concern for off-target application. Later in the growing season, as plants begin to reach a mature size, nozzles should be changed to those that deliver total application volumes of 80 to 100 gallons per acre. One Wisconsin supplier of application equipment is Contree Sprayer & Equipment Company. Work with a dealer to understand and select the correct nozzle for specific equipment. Calibration of equipment is required to ensure proper rates of products are applied.

If there is difficulty in calibrating application equipment, there are several options to consider. Application ground speed can be increased or decreased to adjust overall application rate. If the machinery is over-applying, ground speed should be increased. If it isn't applying enough volume of solution per acre, decreasing ground speed will increase the rate applied. Another option would be to adjust the pressure of the application equipment. Higher pressure reduces droplet size, and it increases coverage and rate applied per acre. This is not always the case and must be continually monitored. Alternatively, reducing pressure decreases droplet size and rate applied per acre.

Burning can occur and needs to be avoided. Cannabis leaves are thin and fragile, like the leaves of soybeans. Our experience indicates that cannabis leaves have even greater susceptibility to crop injury than soybeans. Some simple practical practices can be implemented to reduce the likelihood for crop injury.

First, we recommend starting the application outside of the field itself along the field edge. Allow sufficient spray volume to be applied to this area prior to entering the cannabis field for application. This will clean out any unwanted solution remaining in the lines of the equipment from prior usage. It will also reduce the likelihood for over-application where the machinery enters the field.

Second, continue moving throughout the field, and spray past the cannabis field edge on the other end. Turn off the solution being applied only after being outside of the field itself. There is a small amount of product wasted using these recommendations, but they greatly reduce the likelihood for crop injury. We believe that with high-value crops such as cannabis, wasting a small amount of product is a better outcome than any crop injury.

The final recommendation is to continue through the field if possible. If equipment malfunction occurs during application, simply shut down the pump, and continue to the end of the field. If the machine stops in the middle of the field after malfunction, any spray solution dripping or flowing from the rig can quickly lead to over-application, which will likely result in crop injury. Continuing to the end of the field and correcting the malfunction outside of the field reduces this risk. Do not try to "splice in" the remaining pass that did not receive spray solution after the equipment is fixed. This tends to result in over-application in the area where the malfunction occurred. Instead, continue on to the next adjacent pass of the field. The only change to this recommendation is if the equipment being used has GPS-mapping capabilities. In that instance, growers can be sure they won't over-apply product to the hemp crop because they can utilize the GPS to find the exact location of the malfunction.

Application equipment must be thoroughly cleaned prior to usage. If used equipment is purchased or if the unit itself has been recently used to apply products that do not meet OMRI requirements, there is a high risk that residual pesticides could be applied to the hemp crop. This must be avoided at all costs. Hemp crops that are grown for cannabinoid production need to pass stringent testing protocols for human health-safety reasons. There are more than 400 residual herbicides that the cannabis will be tested for, prior to the crop meeting those standards for human consumption. A failed test results in termination of the hemp biomass. This is a large loss that can easily be avoided.

Hemp is a bioremediator, and as such, it will store environmental contaminants within its biomass. Those contaminants include heavy metals that come from the soil as well as any herbicide that comes in contact with the plant itself. This creates an alarming scenario when adjacent fields are to be applied with herbicide. That could be neighbors' fields or another field on the grower's own farm.

Wind direction, wind speed, and distance from adjacent crops should be monitored to ensure hemp-crop integrity. Certain agricultural herbicides can "mobilize" based on weather. One example of this is DiCamba. This herbicide has been field documented to move 1,000 feet or more if certain weather occurs after application. In this case, if a neighbor applies DiCamba to a neighboring crop within that distance the hemp crop could be at risk.

Foliar applications to hemp during the growing season need to be monitored for overall crop safety. This is mostly due to improper pH level of the tank mix itself. If a grower has experience with "burning" cannabis leaves due to specific tank mixes, as we do, he/she will understand the feeling that there is nothing that can be done after the fact. What's done is done. We did not burn large acreage, usually just edges of fields where application began or where over-application occurred. Many nights' sleep can be lost in this scenario, and it can be easily avoided. We recommend purchasing litmus paper to test pH levels. It is then easy to monitor pH in the tank prior to actual application.

Tank mixes should be acidified, if necessary, to ensure pH is less than 5.5. Buffering pH to less than that threshold can easily be accomplished by adding either citric acid or emulsified liquid fish products. Citric acid is readily available online or from local vendors. In Wisconsin, emulsified fish products can be obtained from Dramm Corporation. We had success using both Dramm One and Dramm S. Dramm S is more affordable and is more liquid in its nature. Therefore, it's less likely to plug nozzle screens on the application rig. If applying Dramm One, consider removing all of the nozzle screens on the rig. Do not remove the main screen before the pump because pump damage may occur. Follow the application rates on the label for similar crops for both citric acid and emulsified fish products.

Applying cannabis-approved products

The Wisconsin Department of Agriculture, Trade, and Consumer Protection has restrictions on what products can be applied to a growing cannabis plant.

NOTE: If residing in a different state, refer to a similar state department for information on these restrictions.

Not only must the product be on DATCP's current approved-products list for cannabis (found online), but also one must take caution to follow the label based on application and usage rates. It's likely that cannabis won't be one of the crops listed on the product's label. Although, products are continually being added to the list and updated.

We recommend finding a similar crop type on the label (factor in planting population, plant type, etc.), and use a middle-of-the range use rate. This is if it's not possible to find use rates from a source with prior application experience such as this handbook itself.

When applying products, take care that application equipment is cleaned to USDA-organic standards. Hemp is a bioremediator and will hold residual herbicides and heavy metals in its biomass. This will likely affect the marketability of the crop.

Equipment

Next, be sure that the equipment has been calibrated to apply a specific rate based on application speed, width, etc. We used rates from 25 to 50 gallons per acre early in the vegetative phase of growth. Those rates were increased to 80 to 100 gallons per acre later in the season when late-season fungicide applications required thorough plant coverage.

Nozzles should be sized for application rates based on distance between nozzles, operating width, and ground speed. Experienced cannabis growers have used select nozzle spacing to specifically apply products directly over the cannabis row when plants are small to reduce product costs.

Lastly, determine use of products based on total application volume per acre, and adjust each product accordingly. Most likely there will be many tank additives per application for efficiency's sake, and jar tests should be made on each specific tank mix.

Organic Materials Review Institute (OMRI)-approved products tend to have less Personal Protective Equipment (PPE) associated with handling on the label, but all safety precautions must be taken. Also, Restriction Entry Interval (REI) is likely to be short with OMRI-approved products, but human laborers are also common in cannabis production, so those REI standards must be followed in totality.

Levels of pH in the tank must be carefully monitored and adjusted prior to application upon cannabis fields to limit crop injury. Cannabis leaves are likely to be burned by high pH tank mixes; we have plenty of first-hand experience with this. Ideally, the tank should be buffered down to a pH of 5.5 with the addition of a dilute acid (for example, citric). Emulsified fish products have the same effect in lowering tank pH as citric acid, so if an emulsified fish product is in the tank the pH will likely be in a safe range.

There has been plenty of "chatter" in the industry that applications of dilute acids to growing plants can result in higher THC concentrations later in the growing season. We did not see this on our production acres, where we applied both citric acid and multiple emulsified fish products to all acres. This "chatter" is similar to that of late-season N applications. Neither is likely to "push" the plant toward THC. Those levels are pre-determined based on genetics. Selecting high ratio genetics gives the grower a larger window to have the crop state-tested for compliance.

CHAPTER 10

Insect Pressure

There are many identifiable insects that feed on cannabis. The majority do not cause economic damage to crops. However, certain species can have populations that negatively affect plant growth and/or yield. Those include: flea beetles, black cutworm, armyworm, aphids, and eurasian hemp borer. The majority of these insects can be controlled using two OMRI-approved products. Use of azadirachtin (active component of neem oil) in a foliar application can control hard-shelled insects. We used it early for flea beetles with great results. Foliar-applied Bt (bacillus thuringiensis) can control lepidopteran species at larval stage.

NOTE: Eurasian hemp borer larvae spend only days on the exterior of the plant after hatching, and then they tunnel inside, where Bt has no effect. Control of eurasian hemp borer with Bt must be managed correctly by field-sweeping adults to determine application timing.

Scouting for insects is a critical portion of any hemp Integrated Pest Management strategy. Another portion of that strategy is the use of pre-plant or drench liquid applications of beneficial bacterium. Those beneficial bacterium then are taken up by the plant's roots and become systemic. Do not expect these to keep your crop free of insect pressure throughout the season.

Keeping the plants well-fed is another important portion of the IPM plan. Plants that are healthy have more natural resistances to insect pressure/feeding. Understanding plant needs and timing of applications is critical to keep the plant growing at or near its maximum potential throughout the year.

Integrated Pest Management plan

1. Keep plants well-fed.
2. Use pre-plant or drench liquid applications of bacterium.
3. Scout for insects and apply products as necessary.

Those early-season drench or pre-plant side-dress applications that include either neem oil or beauveria bassiana can create systemic systems within the plant for defense. Insects must feed on plant tissue and consume enough bacterium to expire. This may be enough to keep populations in check. Specifically, we had success with beauvaria bassiana early, but then we found that the effects wore off mid-season. Another application of beauveria bassiana midway through the year should provide late-season help against third-generation eurasian hemp borer larvae feeding.

Eurasian hemp borer is the insect in Wisconsin that's most likely to cause severe crop damage or complete crop loss to cannabis fields. Populations, if kept unchecked, can grow to more than 20 individuals per plant. Moths lay eggs on the bottom side of leaves. The larvae hatch and spend five to seven days on the exterior of the plant. This is the time when Bt (bacillus thuringiensis) can be an effective application strategy. The larvae then bore into the inside of the stems and branches, usually at node locations. Holes can be visibly seen at these locations.

Once inside the stems, the borers feed and grow through three instar phases prior to pupating. The entire life cycle of the species is about three weeks, and three generations will likely hatch during the season. The third generation is the most damaging because the population grows exponentially, and the larvae have mature bud material to feed directly on from the inside out.

Control of eurasian hemp borer should be a multi-step approach.

Step 1: Keep the plants well-fed to increase the plants' natural defense mechanisms.

Step 2: Two soil-based applications of beauveria bassiana should help with systemic resistance.

Step 3: Use sweep nets when scouting fields to take counts of adults. This creates "application windows" for the use of foliar Bt. Remember, Bt will only eliminate those larvae that are on the exterior of the plant. Once the larvae bore inside they are safe from Bt. If applied on two five-day intervals after the three main generations of moths are present, one could greatly reduce the population. Pictures of adult Eurasian hemp borer moths can be found on Colorado State's hemp website.

Step 4: The final tool in the bag for defense against eurasian hemp borer is promoting the populations of species that directly feed on the borers themselves. Beneficial insects, such as minute pirate bugs, enter the plant through the holes that the borers create, find the larvae, and feed on them directly. If populations of minute pirate bugs are great enough in the field, then a natural defense is created, keeping the cannabis plants free from damaging borer feeding. There are ways to increase the natural population of minute pirate bugs in your fields, including planting bordering plant species that attract the insects. Another way is to purchase live minute pirate bugs, and place them in your fields. Either method will be most effective if started early in the season. Releasing pirate bugs when they have the first generation of borer larvae available to feed on will keep the pirate bug populations increasing. However, releasing them too early with nothing to feed on will result in loss of the beneficial insects. Minute pirate bugs also feed on other soft-bodied insects such as aphids, whose populations can also cause economical damage to cannabis production.

Another beneficial insect to combat eurasian hemp borer is green lacewing fly nymphs. These predatory nymphs actively seek out soft-bodied insects to feed on. If their populations are high enough, they will combat growing populations of eurasian hemp borers. Green lacewing fly eggs or larvae can be purchased live from suppliers and dispersed throughout the field.

A third beneficial way to combat eurasian hemp borer is to increase the population of predatory trichogramma wasps. These tiny wasps actually parasitize the eggs of many moth species, which includes borers. Although they are non-specific predators, trichogramma wasps are extremely effective in parasitizing eggs of other insect species. Live populations can be purchased and dispersed throughout the field.

Aphids are sap-sucking pests in the mite family. Control of the population of these insects will be most economical by increasing the populations of predatory species. We did see large populations of aphids emerge during the hot dry portions of the peak summer months. This is similar to the scouting of other aphid species that affect traditional row-crops. Increasing populations of lady beetles, minute pirate bugs, and green lacewing flies should drastically decrease the populations of aphids in hemp fields.

We did make an application of azadirachtin late in the growing season to see if we could visibly notice a reduction in eurasian hemp borers. We did not see that effect; however, we noticed that the continuing populations of aphids in the field seemed to be greatly reduced following application. We do not make claims as to the direct efficacy of this insecticide against aphids; we are simply noting our own observations from our field grow.

CHAPTER 11

Disease Pressure

Cannabis can be affected by both bacterial and fungal disease pressure. Understanding the causative agents is the first step to keeping plants healthy throughout the year. The next step is understanding the timing of infections and mitigating risks surrounding that timing. Diseases that are of economical relevance to cannabis production include: downy mildew, septoria leaf spot, black helminthosporium, schlerotinia (white mold), bacterial leaf spot, gray mold, and botrytis (bud rot).

The most economical method of avoiding disease in cannabis production is keeping the plants in a healthy state during the growing season. Stressed plants have increased likelihood for disease. Infected plants not only require a treatment-based fungicide, but also they require supplemental nutrients for recovery. Preventative applications of fungicides are highly recommended. Understanding what fungicides to use during certain portions of the season is essential.

Early-season fungicides are intended to prevent vegetative-based disease pressure from occurring. Leaf diseases such as downy mildew and septoria leaf spot can be prevented/suppressed fairly cheaply using a potassium bicarbonate product. These products work to change the pH on the surface of the cannabis leaf, thereby creating an environment unsuitable for disease growth. Most, if applied on a bi-weekly or 10-day interval throughout the season, can be extremely effective.

Scouting is the best way to understand what is happening in the field. New lesions that occur on leaves should be identified and/or sent to a lab for analysis and identification. Any copper-based OMRI-approved fungicide can be used as a rescue treatment for severe fungal infections such as septoria, downy mildew, bacterial leaf spot, or helminthosporium.

Ideally, preventative treatment would be applied near July 1 to keep plants clean and repeated August 1 or just before flowering. Copper is activated when wetted, so any overhead irrigation or rainfall events will further activate the fungicide and kill any active spores. Copper-based fungicides aren't labeled to be applied to flowering plants, so these applications can only occur during the vegetative portion of the growing season.

When treating an infection from a bacterium, a last-resort method for control would be an application of hydrogen peroxide (at least 3%) containing bactericide.

CAUTION - These applications can cause "whitening" of plant tissue and are to be applied only if necessary. Whitened plants have an increased recovery period along with an extra nutrient requirement for vegetative growth.

"Bud rot" is a visual mold inhabiting the cannabis flowers, which is caused by either botrytis or gray mold. Treatment of infection tends to be expensive, and economical loss will likely already have occurred. Preventative measures to avoid botrytis infection are highly economical. Early-season soil-based applications of Serenade Opti (bacillus subtilis), along with a foliar application mid-flowering, can be enough to keep flowers clean until harvest.

We had success with this approach using overhead watering, which increased the moisture in the upright cannabis flowers. Other products such as BotryStop (ulocladium oudemansii) are beneficial agents that inhabit the same space as botrytis, therefore leaving no room for the destructive agents to propagate or sporulate. This specific product is extremely expensive for treatment on a per-acre basis.

A new product recently approved by the U.S. Environmental Protection Agency for use on cannabis is Bacillus Amyloliquefaciens Strain D747. It is another beneficial bio-fungicide that colonizes the root structure of cannabis and works within the plant to help it fight diseases. We have no experience working with the product but may do some experimentation in 2020.

Another recently approved product is Extract of Reynoutria sachalinensis. These products are extracts of giant knotweed. It has been found to increase natural defense pathways in plants when foliar-applied to vegetative tissue. Although we have no experience in the use of this bio-fungicide, it is one of few that are known to be effective when applied in a foliar application as opposed to those that are more effective when soil-applied. This bio-fungicide is best-suited for defense against gray mold and powdery mildew.

OMRI-approved fungicides touting natural fungal defense using natural oils such as peppermint oil, cottonseed oil, clove, palm oil, rosemary, etc. are use-at-your-own-risk products. We used a number of these products with little-to-no efficacy for either prevention or curative fungicide approaches. They simply did not seem to protect from or reduce infection from fungi. Those types of products are also costly and require a short strict reapplication interval. We followed the suggested application procedures with great regret.

CHAPTER 12

Weed Control

Weed control is an essential portion of any agronomic crop plan for row-crop production. Considering cannabis-produced products must be marketed as either organically grown or USDA-certified organic, weed control can become costly and problematic. The use of herbicides to control weeds is prohibited from hemp grown for cannabinoid production. There are multiple approaches to weed control regarding feminized cannabis production, which include: cover crops, mowing, cultivation, mechanical weeding, plastics, and hand labor. Hand labor is the most time consuming and costly.

Cover crops

The use of cover crops can be an essential portion of any weed-control package for hemp production. The species selected to provide cover should not hinder the growth of the hemp plants themselves. Species that are short in stature and do not consume large amounts of macronutrients and micronutrients, which should be left available for the hemp crop, are recommended.

Cover crops can be planted in the fall for the following season's hemp production. Those species that die naturally due to cold temperatures can easily be managed when going into the planting of the hemp crop. Removal of excess residue covering the row itself is then essential for both the transplanting and direct-seeding approaches.

Cover crops that over-winter and continue growth in the spring provide weed control until the time of hemp planting. That's a positive, but the negative component is that live cover would need to be terminated prior to planting, so the cover crop does not out-compete the growing hemp plants. Termination in both models is usually tillage practices. There is one OMRI-approved herbicide approved as a "burn down" of grasses pre-plant. Its trade name is Weed Slayer. It's approved for USDA-certified organic acreage, and it tends to be costly based on usage rate. This would be one option for fall-planted cover crops in the grass family that need to be terminated in spring prior to planting hemp outside of tillage.

Cover crops can also be used in season to suppress weed growth. This can be a useful tool if properly implemented and continuously managed. We broadcasted one bushel per acre of winter rye and one pound per acre of crimson clover pre-plant to all of the acreage we grew in 2019.

Winter rye does not go through a process called vernalisation, or exposure to low temperatures. As a result, spring-planted winter rye does not grow tall like fall-planted winter rye. Instead it grows to a height of 12 to 16 inches, based on our experience, and then falls over. It has low levels of lignin, which is the component within plant stems that hold them upright. Spring-planted winter rye can easily then be mowed between hemp rows throughout the growing season for weed suppression. Row spacing would have to be adjusted based on the size of the mowing unit itself.

Sativa and weed control

Some of the acres we planted in 2019 had heavy giant ragweed pressure, and we cultivated the hemp crop to terminate weeds. This also terminated the cover crop between the rows on those acres. After that we didn't see a benefit in pre-planting the cover crop itself. However, this did not happen on all of our acreage. The sativa varietals we planted grew vigorously, and we had them planted at the correct population to reach canopy when mature. We cultivated those acres and then terminated the cover crop, and the crop out-competed the weed pressure. For this reason, we wouldn't recommend a live cover crop for sativa varietals that can adequately reach full canopy.

Ruderalis and weed control

Ruderalis, or auto-flower, varietals can be successfully managed for weed pressure utilizing cultivation or mowing. Because they will flower after a few weeks of growth and due to their short overall stature, autos can be cultivated multiple times during their short vegetative and flowering phases. Once the auto-flower plants reach a height past that which can be safely cultivated they are most likely already in their reproductive phase (flowering) and will soon be ready for harvest. Weed control for auto-flower varietals only needs to be managed for a short duration of approximately two months, starting from the time they are planted until they are ready for harvest. This short duration makes weed control easier to manage. Mowing can be effective with autos between the rows if the rows are sufficiently spaced.

Indica and weed control

We found that weed control with indica-based varietals can be difficult. That is because indicas are slow-growing, reach a short mature height, and mature late in the growing season.

The first recommendation we have for weed control with indicas is to plant them at the correct density. Planting indicas at rates lower than what is recommended will lead to gaps in the field and spaces for sunlight penetration, even after the plants reach their maximum vegetative growing potential. This is extremely problematic for a crop that will not be ready for harvest until mid- to late October. Weed-control management then lasts from late May until late October, or about five months. That is why indicas must be planted at a proper density to reach canopy and suppress weed growth.

We planted our indica-based varietals at 2,200 plants per acre. This was a mistake. That number should be closer to 3,000 plants per acre or more. Remember that there is loss associated with planting anything, and one must plant extra seed or extra plugs to make up for losses before a desired final stand population is reached.

We found the indicas to be the most challenging. We broadcasted cover-crop seed pre-plant, and that seed germinated and began to grow about one week after transplantation. With the indicas, the rye cover grew faster than the hemp plants. This was a large problem, and we were forced to hand-weed out the cover crop we had planted around our indica plants to allow room for sunlight and growth. This was a costly mistake because labor is not cheap.

After the indica hemp plants had grown past the cover, we then found the cover to be essential in suppressing weeds due to our low overall population. Remember, the rye hadn't vernalized, so it was down on the ground like a mat and suppressed weed growth. We could tell by then that the plants would never reach canopy and that the live cover crop was essential in helping to keep the weeds at bay.

For indicas, we recommended planting at the proper density, so the plants can reach canopy for weed suppression. If low populations of indicas are absolutely necessary, we would recommend broadcasting cover crops such as the mix we planted three to four weeks after transplantation. Indicas need time to "get ahead" of the coming cover crop that's intended to be planted. They grow slowly, so wait until multiple weeks pass by. You may try cultivating once or twice prior to broadcasting the cover crop seeds.

Another option would be to cultivate indicas up until they are too tall to "walk over" with a tractor and cultivator. Then quickly broadcast cover-crop seed in the middle of the growing season to provide weed suppression up until harvest in late October. We did not try either of these approaches, but it seems possible to achieve relative success using those strategies.

Plastic as weed control

The use of plastic as a tool for weed control can be especially successful in suppressing weed growth. This is a common practice with other specialty crops and must be accomplished at planting. There is added labor, time, and cost upfront, but management in-season is greatly reduced. Success with weed suppression is quite high. We would recommend this approach for any growers who have the capability at planting. Modern transplanters and some direct-seeding planters can accomplish this feat. However, time, labor, and management may be too great for large grows. Keep in mind that plastics must be picked up at the end of the season, which is additional time, labor, and expense.

We do know multiple growers who used a bioplastic that does not need to be picked up at the end of the season. This plastic was laid down with a direct-seeding planter. The response from those growers was the same: weeds grew right through the plastic itself in season. Not only is that added cost upfront, but also if you have to continually weed after the plastic is laid (as if it was not even there), that is especially costly. For that reason, we do not recommend using bioplastics for successful weed control.

Hand labor for weed control

Hand labor is one way of keeping hemp fields clean. This is the most labor-intensive and time-consuming option. Time is money, and even if you are doing the work yourself, you must factor in the expense of your own labor. Hand-weeding can be extremely costly. We hired professional H-2A labor for portions of our weed control in 2019. We also did plenty of weeding ourselves. Our average cost was $1,000 or more per acre. For small acreage it can be managed, but again, sativas or autoflowers would be recommended rather than indicas. That's because the autos only need to be managed for two months, and sativas, if planted at the correct density, will reach full canopy, which chokes out weed growth. Hand-weeding indicas is likely five months of work unless large indica varieties are selected and population density is adequate.

Mowing as weed control

We have mentioned mowing throughout this chapter as a possible solution for weed control. It is best utilized on small to medium acreage, which is about three to 15 acres. Row width must be sufficient to accommodate the specific mower that will be used. Mowing will not eliminate weeds between plants within the row, but if implemented on a bi-weekly basis, it can be extremely effective. An ATV along with a gas-powered hand-held trimmer or weed-eater can then be taken down the rows and used to eliminate weeds between plants.

Mechanical weeders

The final option we mention for weed control is using mechanical weeders. Some small units are available for specialty crops that have rotary-turning finger-style tines on each side of the crop row. Most are built as either pull-type or three-point mounted units on the tractor itself. They require one person as the operator and another person on each row unit. The person on the row unit can manually move the rotary tines closer to or farther away from the row during operation.

A Company called A1 Implements has created the Hemp Hawk mechanical weeder based off the design of these older specialty crop units. They are commercially available.

We envisioned combining this older technology with another basic unit, namely a self-propelled "Hiboy-style" chassis. Hagie 284 units that are 1990s-models or older are fairly cheap pieces of equipment. The cost lies in creating this system. Growers would have to purchase the two pieces of older equipment and then fabricate both units into one. A professional with a high mechanical skill set is required for this final step to create the custom-made mechanical hemp weeder. We envisioned using one of these old Hagie models, without an operator's cab, to reduce cost (get an umbrella to keep the sun off you), along with a four-row rotary-tine weeder mounted on the back. Five people would be required to operate the unit, but once made it would likely be extremely effective for the entire growing season. Bi-weekly passes across large acreage could keep fields free of weeds at relatively low cost per acre. This would include the cost of building the rig, and the labor of five people to operate it during the growing season.

CHAPTER 13

Irrigation

Overhead center-pivot irrigation systems

Modern center-pivot irrigation systems utilize overhead pipes that arch from tower to tower with spans of several pipes bolted together within a suspension frame. Each tower consists of two wheels that follow each other in a circle-shaped track around the field. Attached to the two wheels is a triangle-shaped steel frame, with the base of the triangle at each wheel. The top tip of the frame connects to both the pipe-suspension frame in front and behind the tower.

The full length of towers and spans are determined by the size and shape of the field. In some fields, only a one-quarter, one-half, or three-quarter turn can be made, due to field shape and/or size. However, most center-pivot irrigation systems travel in a full circle. The whole system is attached to a center-pivot point in the middle of the field, which is anchored by a concrete pad.

The speed of the tower drive shafts (driven by electric motors) controls the speed of each tower's travel. That in turn controls the amount of water applied to the field. The towers farther away from the center-pivot point travel faster because they need to cover more distance out at the edge of the circle. Towers closest to the center-pivot point travel slower and less distance. Heavy hose joints with internal hook and pin joints at each tower tie the system together and allow the system to flex as each tower advances. Tower motors are wired to turn the transmission in both directions, which means the whole system can travel in forward or reverse directions. That is important in fields that are not a full circle and for moving the system out of the way for field operations.

The ease and flexibility of a center-pivot irrigation system allows hemp to be grown on sandy soils. Use of the system in the spring at planting time is important for keeping the top three inches of the soil moist during germination and emergence. Sandy soils can dry within a matter of a few hours when wind speeds increase during the day.

Center-pivot fertigation can also be effective and efficient for growing hemp. Any substance that is soluble in water and is taken up through the hemp root system can be applied through the irrigation water. Injection pumps and tanks located at the center-pivot point hold and inject nutrients into the irrigation stream. Pumps are calibrated to flow rates that synchronize with the amount of water applied to control the rate of nutrients applied per acre.

As plants grow and the hemp canopy becomes more closed, moisture on plant leaves, stems, and flowers tends to stay longer in the morning. That moisture can be from dew, rainfall, and/or irrigation.

In July and August, hemp will shift from vegetative growth to both vegetative and reproductive growth. When flowers start to form and become sticky as they attempt to capture pollen, those flowers become susceptible to both gray and white mold diseases. At that point, it is important to shift to a early daytime irrigation application, so the flowers dry as soon as possible after the water is applied. Nighttime irrigation is more efficient in terms of less evaporative losses due to cooler temperature, less wind, and no sunshine. However, the risk of initiating and aggravating levels of gray and white molds from a long overnight wetness period is too great.

Traveling gun irrigation systems

Traveling gun systems utilize on-the-surface rolled-out hose. They were developed for small and/or odd-shaped fields or enterprises with limited funds. These systems utilize an approximate six-foot vertical aluminum pipe attached to a rotating four-foot nozzle at the top. Growers can choose different nozzle fittings for the tip of the hose in order to help regulate water flow and spray distance.

A mounted wand or weighted apparatus near the tip oscillates like a large grandfather clock pendulum to intercept the discharged water stream. That stream breakup serves to make the irrigation water fall in a more uniform cloud pattern throughout the distance that the water is traveling. The irrigation riser and gun with nozzle are mounted to a four-wheeled cart that travels down pre-planned lanes within a field. The car is placed at one end of the field to start a run up the lane. Water is piped above ground or from a riser within the lane. A flexible rubber hose is attached to the end of the solid pipe or to the permanent riser. Pipe or risers are fed irrigation water from a nearby irrigation well. The rubber hose is laid out on the lane from a reel on the cart.

A turbine is built into the base of the cart, which is where the water passes on its way up the cart riser. The blades on that turbine are turned by the passing water to drive a spool, where heavy cable is wound. Once the cart is placed at the start of a run, the cart is disconnected from the towing vehicle (usually a tractor). The towing vehicle is then driven to the opposite end of the run while pulling the cable with it as it unwinds from the spool. The vehicle is parked at the opposite end with the cable still attached. The vehicle serves as an anchor to hold the cable as the water drives the spool to wind up the cable.

Thus, the traveling gun irrigates the field as it travels slowly toward the anchor vehicle as the cable winds up on the spool. The amount of irrigation water applied to the field is controlled by a transmission connected to the turbine. That transmission adjusts the speed the cart travels at, which thereby controls the amount of irrigation water that is applied.

Fertigation can also be accomplished with a traveling gun because the nutrients to be applied are injected into the water stream at the well. All irrigation wells must be equipped with a backflow-prevention device built into the main discharge pipe about two to three feet from the well itself. That backflow-prevention device is necessary to stop nutrients from being siphoned back down the well when the well pump is either purposely or inadvertently shut down.

Because a traveling-gun irrigation system is an over-the-top irrigator like a center-pivot system, the same precautions regarding gray and white disease-cycle prevention are in order. Nighttime applications of water are easier with a traveling gun due to less wind velocity at night. Wind has a greater adverse effect on the single water stream on a traveling gun versus the many individual nozzles on a center-pivot irrigator. Even though that is the case, daytime irrigations with a traveling gun should be made during the flowering period for hemp.

Dripline irrigation

As previously mentioned, we have little experience with dripline irrigation systems, but we understand their use and importance for specific grows. Feminized hemp grows varying in size from one to 10 acres are best-suited for dripline irrigation systems. A low pressure water source is required, along with laying poly-water line at-planting to "drip" feed the plants throughout the year. Applications of liquid nutrients are especially user-friendly with this approach, and it has been targeted by most first-time or small-acre growers.

Dripline irrigation is not recommended for dual-purpose hemp production.

CHAPTER 14

Male Removal

Dates according to varietal

In Wisconsin, male identification won't need to be considered until about the second week of July; this date applies to sativas and indicas. Early-planted autoflower populations would be an exception to that general date. Autoflower fields should be continually scouted and combed for males one to two weeks prior to female presentation.

Identifying males vs. females

Females will present pistols that appear to be long thin spines or spikes near the top of the plant, which is where flower formation will occur. Males do not present pistols, and therefore, the absence of pistols is a precursor to pollen sac production.

Males can easily be identified by pollen sacs that are hanging throughout the plant. We noticed that the earliest-forming sacs tend to be located near the top of the male plant. That area of the plant should be scouted in detail to identify and remove males as early as possible.

Scouting and screening

Fields need to be screened periodically to check for males.

In order to efficiently accomplish scouting fields, we bought a small motorbike that could be ridden between the rows. This worked fairly well; although, you need a good limber operator to drive down the row, scout using peripheral vision, and carry a loppers all at the same time.

We effectively removed males across 60 acres using this methodology. The remaining 30 acres we had in production were feminized less desirably. By that, we mean the seeds for these acres were purchased from less reputable breeders, and likely, the feminized seeds themselves were grown outdoors. Any feminized seed grown outdoors runs the risk of pollen contamination from neighboring plants. On those acres, we saw greater male percentages, and we were forced to periodically walk those acres with small crews to remove all males. In either situation, it's essential to check each row by some means to avoid any crop loss from pollination.

Even if the entire population of plants is 100% female, male scouting will need to be applied. This is because a small number of female plants will likely become hermaphroditic in nature. We have personally witnessed some of these plants present pistols, begin to produce small flowers, and then produce pollen sacs. It's likely that this pollen would produce feminized seeds because there aren't any "Y" chromosomes in the plants. The problem is that any neighboring female plants that are pollinated will immediately change to seed production. Female plants that are pollinated mobilize their energy toward producing seeds. They abandon continued flower production because they have found what they have been searching for … pollen.

Female plants

Female cannabis plants produce flowers. Those flowers are designed to attract and hold pollen to produce seeds for reproduction. If pollen is not present in the surrounding environment, the female plants will continue to produce more flowers. They then, in turn, will produce more cannabinoid content within. This is an outline of the general concept for field production of cannabinoids. The highest-yielding fields are those that are solely female plants. This is easily attained using clones.

Clones are simply propagations of "mother plants," which are identified as female. Cuttings are taken from the mother plant, placed into their own pots with medium, and rooting hormone is added to create a new live plant separate from the mother plant.

Feminized seeds are different from clones because high-quality feminized seeds are considered 95% female. That leaves 5% of the seeds as males within the general population. More recently, this number is close to or less than 1% males. That's due to the use of modern technology within the breeding program.

If a grower uses 2,000 plants per acre, then 1% would equal 20 male plants per acre. In our 2019 crop, we scouted and pulled males across 80 feminized acres. We believe even less than 1% of those 80 feminized acres had to be removed. If 10 acres were in production, then 200 plants would have to be identified, cut, moved out of the field, and destroyed. Some genetics have proven feminization rates at more than 99.8%. That's only four males per acre for a standard seeding rate of 2,000 plants per acre.

Removal of male plants from the field

Males that are already producing pollen when they are identified should be bagged where they are in the field, and then cut and removed. If a male plant that is producing pollen is cut before it is bagged, the act of that plant simply falling over will produce a cloud of fine pollen particles that will pollinate surrounding females.

On the contrary, if the male plant has begun hanging pollen sacs but hasn't yet produced any pollen, there is less risk in cutting without bagging first. Males should be taken to the field edge and collected there. We put our males in large bags, put the bags on trailers, and hauled the biomass miles away from the field for disposal.

Some growers have made small steel incinerators on the edges of their fields for destruction of male plants. This makes a lot of practical sense for male removal. To make one of these, we recommend purchasing or fabricating an old small dumpster with a steel lid. Throw the males inside, and burn. However, the unit may need to be altered in order to draw airflow to keep the fire burning.

Cannabis pollen is engineered to cling to anything it comes in contact with (for dispersal reasons), and cannabis plants are wind-pollinated. However, it's possible for plants to become pollinated in other ways. You can cut down a pollen-producing male plant, acquire traces of pollen on your clothes, and thus spread pollen across the feminized field for the rest of the day. Care must be taken to minimize this. That's why it is especially important to bag pollen-producing males right where they stand prior to cutting and removing them.

CHAPTER 15

Harvest

Harvesting mature feminized hemp plants will likely be the most challenging and important step in ensuring commercially viable marketable material. We highly recommend varying planting dates and planting specific varietals to lengthen the harvest window for management reasons.

Large acreage should utilize all three types of cannabis to ensure a successful harvest season. Autoflower varietals can be planted late in the planting season. These varietals will mature in August and be the first that need to be harvested into a stable state. Sativas will reach maturity throughout the month of September and into early October; a certain portion of acreage should be planted to utilize this window in the harvest season. Indicas should be planted early to maximize their vegetative potential. They will mature late in the year and can be harvested into the first week in November. CBD concentration can be affected by freezing temperatures, and avoiding this situation is recommended.

Hand harvest

Choosing a harvesting method depends on the amount of acreage and the cost a grower is willing to take on to successfully manage the crop. For small acreage, hand harvest is an option. Hand harvest entails cutting the plants off at their base, transporting them from the field to a drying facility, and drying the entire plant down to less than 10% moisture.

Live plants contain approximately 70% moisture, the majority of which needs to be removed, so the remaining biomass material is in a stable state. Hanging plants in buildings with adequate airflow, using tobacco-drying flues or "long barns," or mechanical forced-air dryers are all options for this harvest approach.

Plants must be dry in the field before harvest can commence. Harvesting plants with dew still remaining on the surface of the plants results in increased risk associated with fermentation development. All plants that are loaded into transport vehicles will begin to "heat" after a certain period of time (sometimes a few hours). Wet plants loaded in this fashion quickens the fermentation process dramatically and should be avoided.

Cutting plants off at their bases and loading them into transport wagons is a labor-intensive time-consuming process. A loppers, or large pruning shear, can be utilized for this practice. Motorized trimmers with steel cutting blades are another option that increases efficiency.

A crew of laborers is required to cut the plants, load them onto transport vehicles, and move them to the drying facility. If hang-drying the plants is the determined drying method, then another crew of laborers is required to hang all of the plants that were cut that day.

Mechanical harvest

Mechanical harvest greatly increases the number of acres that can be successfully harvested in one day. This type of harvest has multiple options including: whole/partial plant, stripper header, and Case New Holland (CNH) combine rotor. Understanding how the material will be dried post-harvest will determine which method to use. Whole plant material can be hang-dried, dried in tobacco flues, or shredded and dried with a forced-air dryer. Partial plant material can be dried in tobacco flues, or shredded and dried in a forced air dryer, rotary dryer, or kiln.

Whole-plant harvesters exist that cut the plant off at its base and load the plants onto wagons for transport. One on the market is made by Kirby Manufacturing. It is a three-point tractor-mounted piece of equipment that requires one operator and another laborer stacking plants in the transport wagon. These green whole plants can then be dried using hanging, tobacco flues, or shredded to be dried in forced air dryers, kilns, or rotary dryers.

Use of a mechanical stripper header or modified combine rotor to harvest green mature hemp plants results in green bud/leaf material with most of the unnecessary stalk already discarded. This saves steps in the harvest process and increases efficiency. Drying that type of material can be accomplished using forced air dryers, rotary dryers, or kilns. Wet buds harvested in this manner must be dried immediately to avoid fermentation and associated losses.

Stripper headers actually "strip" the green buds off the remaining stalk/fiber structure of the plant and leave that portion of the plant in the field. Some growers have mounted these heads onto forage harvesters, which chop or grind the material. We recommend avoiding this methodology in totality.

Trichomes are fine plant parts within the flowers themselves that produce oil. They are the portion of the flower that houses the largest amount of cannabinoid production. Any chopping, grinding, or milling of plant material containing trichomes should have a fully enclosed dust-collection system in place to capture those trichomes that become airborne. Significant losses can result from utilizing a stripper header mounted onto a forage harvester. Use of an Oxbow chassis, which is a specially designed piece of machinery for specialty crop harvest, can be extremely efficient harvesting large acreage with a stripper header. There is significant cost with the equipment of this harvest method.

CNH-designed combine rotors can be modified within the existing chassis itself to separate green bud material from stalk. These combines are equipped with small-grains harvesting heads that cut the plant branches using a sickle and feed the material internally into the combine. There are "kits" currently on the market by various manufacturers to transform a CNH combine designed to harvest dry grains into one that is capable of harvesting and separating green bud material for hemp. The remaining wet bud/leaf material must be dried immediately. Drying of such material can be done using forced-air dryers, rotary dryers, or kiln drying. Internal dust-collection systems are advised on all mechanical equipment intended to harvest high-cannabinoid content bud material.

A final option for successfully harvesting feminized hemp plants is building custom-made harvesting equipment. We executed this final option and successfully harvested not only the full 80 acres of feminized hemp that passed testing, but also the 35 acres of dual-purpose hemp that was grown experimentally.

We re-designed a swather head, which was originally designed for small grains. This custom head was mounted onto a tractor equipped with a front-end loader to carry, as well as raise and lower, the head itself. We designed the head to convey the cut plant material to one side and discharge out to bottom. We then mounted a side/rear discharge conveyor on the bottom side of this head. The conveyor was mounted to carry the material vertically to the rear of the loader tractor and "dump" the material onto a trailing forage wagon. We built two of these rigs, so we would have a back-up if breakdowns/catastrophe ensued. With some limited mechanical fixes in season, we were able to harvest our material and then transport it to our drying facility. Our rig required two to three people to operate for feminized-hemp harvest, but it only required a single operator to harvest dual-purpose hemp.

Compaction baling

Use compaction baling at grower's own risk

Compaction baling wet hemp material is one way to store the material for physical transport. It in no way ensures that the material itself is in a stable state. It does not remove moisture from wet plant material.

Compaction balers use pressure to bale plant material into a round bale that is then wrapped in plastic. Many growers harvested their wet hemp material in 2019 and compaction-baled it thinking that they had successfully harvested the crop. The problem with this method is fermentation. Fermentation exists in an anaerobic environment that contains moisture. That is exactly the environment created inside a compaction bale. The only way to avoid fermentation is to decrease the internal temperature of the compaction bale itself from 70 degrees Fahrenheit to less than 40 degrees.

We did experiments using our own compaction baler to prove this point correct. We found that applications of liquid nitrogen to the bale as it is growing in the baling chamber would decrease its overall temperature. The problem is that it quickly becomes uneconomical, due to the cost of liquid nitrogen, to apply enough nitrogen to the bale to change its temperature significantly.

If compaction baling is to be used to handle wet hemp material, then it should be used with strict guidelines. The hemp could be immediately dried. In this case, compaction-baled materials, would avoid fermentation because water is immediately removed, creating a stable material state. If compaction bales are to be stored before drying, then the internal temperature of the bale must be decreased immediately after baling to avoid fermentation.

This can be accomplished by housing compaction bales of wet hemp material in cold-storage facilities immediately after baling occurs. Fermentation will occur within a few hours, and that needs to be avoided entirely. Bales can be held in cold storage until dryer availability is secured. The bales still need to be broken down and dried, which can only be accomplished with forced-air or rotary dryers.

Fermented compaction-baled hemp material smells. The smell is the odor of gases produced by the bacteria that drive fermentation. This smell carries with the material through the processing procedure. CBD-oil distillate that is produced from hemp biomass that has been fermented will also contain this smell, which is unwanted by consumers. Many growers believe that the smelly distillate oil can be "isolated" into a dry crystalized CBD powder, which is still a commercially viable product. We found, through further research, that even isolating CBD from fermented compaction bales "smells like feet." We recommend avoiding this.

Testing during harvest

As the hemp crop approaches harvest, one needs to monitor the cannabinoid level of the crop for CBD and THC (and all their derivatives). In states that are following federal guidelines, those analyses must be performed at an FDA-approved laboratory. However, the crop needs to be monitored more frequently than any governmental agency does, so growers know when to have their hemp crop officially tested. In addition, self-monitoring will indicate how well the crop is responding to fertilizing, watering, and pest-management programs in-season.

In 2019, Driftless Extracts LLC utilized Rock River Labs from Watertown, Wisconsin, for in-season cannabinoid analysis. The lab produced timely results. Two to four days after samples were delivered to them, results were emailed to us. Their fees for analysis were competitive as well.

Samples were collected early in the morning and driven directly to the lab to further reduce the time between sampling and availability of results. Sampling for cannabinoid levels in hemp should be done with a stainless steel shears if one is analyzing for heavy metals. Take the top two inches of the main hemp cola, and collect about 25 of those plant tips. Keep samples cool by placing them in an ice chest until they are delivered for analysis.

CHAPTER 16

Transporting Hemp

Harvested hemp biomass materials can be transported in a number of different ways. The proximity to the drying facility will determine what type of transport vehicle is necessary. For acreage with drying capability in close proximity, wagons of varying sorts can be used to move the material. These wagons need to be thoroughly cleaned in order to ensure an end product that is free of unwanted contaminants. Wagons that can be unloaded themselves are more efficient at moving the material than those that need to be manually unloaded. That basic principle holds true for all transport methods that could be used.

We used forage wagons that were open-topped to move our material from the field to the drying facility. Those wagons were equipped with hydraulically driven aprons that unloaded the material out of the back of the wagons onto concrete pads at the drying facility. This type of transport mechanism is efficient for production of large acreage where drying facilities are nearby to field grows. We successfully harvested and transported 80 acres of feminized hemp using three of these wagons. They were able to move material away from our custom harvester that was clearing up to 10 acres per day.

Other methods of transport may need to be smaller, rather than larger, to accommodate specific specialty markets. Smokable-flower harvest, for example, would take the top portion of the plant only and leave the remaining portion intact. We accomplished this feat by leaving "harvest alleys" or complete blank rows on intervals throughout the field. It gave us access to the middle of the field for scouting, specific weeding, male removal, and smokable-flower harvest. This gap in the field was only eight feet wide, so a specialized piece of equipment was necessary to maneuver down these alleys. We chose a 40-horsepower Ford loader tractor that pulled steel Apache livestock bunks with wheels to harvest the smokable flowers. These units moved right down the harvest alleys and allowed us to harvest the unplanned 40% auto-flower that presented itself from two different strains while leaving the remaining plants intact.

If transport is necessary for large acreage or across large distances we recommend utilizing tractor-trailer semis for transport. No matter the type of material harvested, specific types of trailers are better-suited for hemp. Those that have "walking bottoms," "live bottoms," or are end-dump are best-suited to move the wet plant material and unload themselves on-site at the processing location. If dry plant material is to be transported, we recommend using the same style trailers, ensuring they are securely covered.

NOTE: All equipment that will be used to harvest and transport hemp intended for human consumption must be thoroughly cleaned prior to usage to eliminate possible contaminants.

CHAPTER 17

Drying Hemp

Different methods

There are many different methods of drying wet hemp plants. The basic concept is to lower the moisture content within the biomass from 70% moisture to at or below 10% moisture. This can be accomplished by plant hanging, tobacco flues, forced air dryers, or kiln/floor drying. We will cover the execution of all of these possibilities, but it is up to the grower to determine the amount of drying necessary to accomplish this feat in a timely manner throughout the harvest season.

Spreading out acreage with specific varietals that mature at different times between August and October is highly recommended to ensure proper time management and overall success. If a grower determines they are going to use off-site drying via a local processor, communication and flexibility is paramount to harvest around variable weather patterns.

Hemp buds that are exposed to high temperatures during the drying stages are at risk for degradation/loss of the cannabinoid content itself. For this reason, we believe that feminized hemp should not be exposed to temperatures more than 125 degrees Fahrenheit. Drying at less than 100 degrees would be the ideal situation to ensure that material is not negatively affected. The temperature and amount of air that the plants are exposed to or that is pushed through the hemp biomass itself will determine the amount of time it takes to effectively dry the hemp.

Hanging

Hang-drying whole hemp plants inside well-ventilated buildings can be a successful drying strategy. Time and labor constraints as well as the amount of material that will fit inside the determined space are limiting factors to acreage.

High-ceiling buildings with adjustable or opening sides or ends are preferable to closed facilities. Hang-drying requires air flow, and a closed facility doesn't lend itself to proper air flow. Cables or high-tensile wire can then be strung from either side-to-side or end-to-end within the building, which will create a lattice to hang plants on. In certain buildings, this lattice may be built vertically to accommodate more plants.

Plants should be hung upside down from one large branch that is capable of holding the entire plant's wet weight. Proper anchor points or support structures may be necessary to ensure the combined weight of all plants will not negatively affect the integrity of the building structure itself.

Plants should be hung close enough to maximize the available space but not so close that they compress each other. Plants packed too tightly have a much greater risk for mold and/or bacteria developing during the drying period.

Use of large fans to draw air across the entire space is recommended. Those fans should all be moving air in the same direction from a closed end of the building. That closed end of the building should have louvers, or slats, that open and close, which will allow new air to enter the building.

Open-sided buildings allow prevailing winds to enter the space and move air across the drying plants. Once filled, buildings should be continually monitored to ensure proper air flow. This can be adjusted during adverse weather. Hang-drying hemp plants is extremely labor-intensive. A crew of laborers is necessary to hang all plants that have been harvested throughout the day. This, along with the amount of total material that will fit inside the building, creates limits to acreage sizes that will be effectively managed during the harvest season.

We hang-dried approximately five acres of feminized hemp in 2019. We found it to be hugely labor-intensive throughout the process. The other constraint is effectively lowering the moisture content within the plants in a timely manner. Once the building is full it could take anywhere from seven to 14 days to lower the moisture percentage enough, so the remaining material is in a stable state.

Proper care must be taken when removing dry plants from their environment. We found that the buds themselves act like sponges. On days of high air humidity the buds that were previously at less than 10% moisture will take in moisture from the surrounding air. This can be extremely problematic if you intend to execute further processing with the plants and store them in a packaged state. Some of the material we tested had as much as 30% moisture when plants were removed on days with high humidity. That material then had to be further dried before processing to ensure safe storage. This added drying is costly.

Tobacco flues

Tobacco flues or "long barns" are enclosed forced air-drying units. Most originate from manufacturers located in and near North Carolina, where expansive tobacco processing has occurred for decades.

To begin with, each barn must be equipped with electricity and some type of gas as an energy source. The burners inside the barns must be properly maintained and vented for safe operations. Inside the galvanized metal buildings are crates that roll in and out during the season. These crates can be moved with specialized equipment capable of handling the expected weight in a safe operating manner. Once removed, each crate can be lowered onto its side and opened to fill with material.

Laborers are required to pack the edges of the crates evenly to ensure proper drying. The material can be compressed using human weight (jumping) similar to grape stomping — only without all the juice!

Once the crate is filled, close the doors. Then push metal rods through holes in the doors and through the plant material itself to hold the material in place when the crate is lifted upright.

Next, stand the crate up, carry it to the barn, and roll the crate inside. The fully loaded crate will weigh as much as 1,800 pounds, so safe transport of crates must be planned for. Each barn contains anywhere from eight to 12 crates. Once filled, the blower and burner can be turned on, and temperature can be adjusted as necessary. Take caution to adjust the damper on the barn, so the barn has adequate air flow. This damper may need to be adjusted if air temperature changes throughout the harvest season.

We used 11 long barns to successfully dry more than 150 acres of feminized hemp in 2019. It can be a useful way to dry hemp at low temperatures. Tobacco flues are labor-intensive to operate. A crew of laborers is necessary to achieve effective throughput and maximize the drying potential of the flues. Proper safety precautions must be taken, and general SOPs should be created to ensure proper technique.

We found that each barn will hold approximately one acre of mature hemp plants. Once loaded and turned on, the barn will take at least three days to dry the hemp from 70% moisture to less than 10%. The barns can then be turned off, crates can be removed, rods taken out (be sure each one is accounted for), and emptied. The dry hemp material can then be moved on to further necessary processing.

Forced air drying

Forced air drying of wet hemp plant material may be necessary for certain acreage requirements. Run times for forced air-dryer capacities and through-put must be taken into consideration for larger acreage that needs to be dried. Those acres should spread out specific varietals planted to have acreage available for harvest, beginning Aug. 10 with autoflowers and ending late October with indicas. This will increase the total harvest season, so all of the acreage does not mature at once and therefore need to be harvested and dried within a short time span.

Plant-material specifics (whole plant vs. branched vs. green bud) will vary the drying time, and therefore they will vary the throughput associated with forced air drying. Most continuous forced air dryers can only handle ground green plant material. There are some capable of handling whole bud material, and we purchased one of these.

Growers should schedule their intended drying times early with processors. These times will have to be adjusted in season based on weather, so constant communication is necessary between the grower and processor to ensure successful harvest of the entire crop.

Forced air hemp dryers are designed to pull moisture out of green hemp material and dry the material to less than 10% moisture. They are designed to be continuously fed, so a constant flow of dry material is produced while running. The continuous drying is the main positive aspect associated with these drying systems vs. other batch-type systems, which require loading and unloading of material between batches.

From there, the hemp material can be further processed and bagged to create a marketable product, or it can be held in cold-storage facilities awaiting run times for extraction to distillate and isolate products. Forced air dryers can be expensive to operate, but they require only a small percentage of the manual labor required for other drying types.

Kiln/floor drying

Kiln drying, or floor drying, is the final drying system we will discuss.

These systems require the hemp material to be loaded into and removed from the system in batches. The kilns are enclosed systems that have heated floors or other heat sources to dry the wet hemp material. Some systems require the hemp material to be "turned over," or mixed, during the drying phase to create even drying based on depth of material across the floor.

Kiln drying does not require high temperatures to adequately dry hemp material. For this reason, these drying types take especially good care of hemp material that should not be exposed to temperatures more than 125 degrees Fahrenheit. The size of the unit itself will determine the throughput, or amount of hemp that can be effectively dried in a certain amount of time. The ease of filling and emptying the kiln/floor unit as well as the time it takes between batches are the determining factors to consider when understanding the specific capabilities of each system.

Dried plant material that will be further processed should be handled carefully. Once dried, the trichomes, flower, and leaves become brittle. Those are the parts of the hemp plant that contain the highest contents of cannabinoids. If those parts are packed, pressed, milled, or smashed with equipment that doesn't have any contained dust collection on it, there will be corresponding losses. Dried material should be handled with extra care.

CHAPTER 18

Dual-purpose hemp

The majority of the previously mentioned practical approaches to hemp cultivation carry over from the feminized model to the dual-purpose model. Specific differences exist, and in this chapter we will discuss those that are directly specific to traditional industrial hemp production. Topics such as insect and disease pressure are the same in both models and will not be overlapped in content coverage.

Seed selection

Industrial hemp can be grown as a dual-purpose crop. In one case, hemp can be grown for CBD and fiber. Alternatively, other varieties can be grown for grain and fiber. Those varieties grown for CBD and fiber tend to be tall, growing to more than nine feet. European fiber varieties would follow that growth habit. Examples would be the varieties Futura 75, CFX-1, and CFX-2 which are grown throughout Europe for CBD and fiber.

Some of those same European varieties would be monoecious in reproduction, meaning that the same plant has both male and female flowers on it. Dioecious varieties would have separate male and female plants. Monecious varieties tend to yield the highest grain amounts when grown for that purpose.

It is highly desirable to utilize short varieties when growing hemp for grain. Combine platforms have a limit to the height they can reach for grain harvest. Varieties that grow only to five to six feet tall are desirable for that grain production. The variety X-59 has that growth habit while producing a nutty-flavored seed for human consumption.

Field selection

Dual-purpose hemp crops prefer well-drained loamy to sandy loam soils, with a cation exchange capacity (CEC) in the range of 4 to 25. In the case of varieties grown exclusively for fiber, irrigated sandy soils make an excellent field site for growing tall plants. Because such crops are harvested before seeds are formed, head diseases of hemp are not a concern.

Because dual-purpose varieties are grown in close-row culture, they are usually planted with grain drills or other close-row planters. Think six- to 12-inch spacing. Fields that are on top of ridges or hills make ideal sites for that type of growth pattern. Early morning winds on those high elevations will dry the evening dews quickly, which help to ward off the onset of head diseases such as white mold and gray mold. That is especially important for hemp fields grown for grain in order to minimize head diseases as the grain matures.

Seeding

Drills equipped with press wheels and depth-band gauge wheels are preferred for direct-seeding hemp in narrow seven- to 10-inch rows. However, European planters with narrow 12-inch spacings are available. They have vacuum seed disks that can plant hemp seeds one inch apart in the row. Those planters may offer the ability to use slightly less seed while still maintaining uniform plant height and stem diameter.

Row spacings of six to seven inches have the advantage of closing the canopy quickly to choke off weed growth and conserve moisture. The disadvantage of that spacing in a hemp crop grown for grain may be lodging, where thin plant stems produced by in-row competition may fall over in a heavy rain/wind storm as their head-grain weight increases.

New European planters with 12-inch spacing are likely to help grain-hemp varieties produce a more uniform plant type with less lodging potential. However, weed control may be a bigger challenge because 12-inch rows require more time to close the canopy.

Field preparation

A firm granular level seedbed is highly recommended for direct seeding hemp for dual purpose. A firm level seedbed is more likely to allow uniform seed placement because row-to-row and within-row movement of seed furrow disks is minimized (both up/down and side-to-side oscillation). Seedbed firmness can be achieved by both rolling the seedbed before and after planting.

Hemp varieties grown for fiber and low-concentration CBD can be managed well on sandy irrigated soils. Because that crop is harvested at the onset of flowering, head diseases are of minimal concern. Therefore, evening and night irrigations are a good management practice in the early to middle portions of the growing season. Overhead irrigation systems work best for any fiber hemp production. Dripline systems don't apply enough water in the root zone of narrow-planted hemp because one could not afford that narrow dripline configuration. Additionally, dripline close to the surface would be torn apart by field operations that cut, rake, and bale fiber hemp.

Any soluble fertilizer can be applied through overhead irrigation systems. Because dual-purpose hemp is planted in narrow rows, it is virtually impossible to pass through those fields with ground equipment after row closure. Therefore, mid to late-season fertilizer applications would be useful when applied via fertigation. Applications of nitrogen and potassium are likely to be needed later in the growing season, especially for hemp grown on sandy soils.

Soil and plant analysis

As in any crop, preseason soil analysis is needed to determine the baseline fertility that may or not be present in any field. But, there is one nutrient to pay special attention to when growing hemp for seed. Because phosphorus (P) is so important during grain formation, P availability throughout the growing season is critical. In addition, any shortage of P during any part of the growing season will lead to a delay in maturity of the hemp seed/grain crop. Frost hazard at the end of the growing season could then be greater. The same could be said of soil pH. When soil pH is less than 6.0, crop maturity is delayed. The lower the pH is from 6.0, the greater the delay in maturity.

Fast-growing hemp for fiber crops should have plant sampling done early in the growing season, perhaps as early as 12 inches in height. That fiber crop will soon be growing at a rate where any nutrient deficiency will permanently retard the eventual maximum crop height.

Slow-growing hemp varieties for grain should be sampled just before entering reproductive growth. P availability will be especially important from that time until maturity. A plant analysis at that time would verify whether or not P levels are adequate to mature the seed/grain hemp to its full potential.

Weed control

Seedbed preparation is critical for obtaining high germination rates and uniform hemp emergence. Thick hemp crops produced by high germination rates will have the ability to choke out weeds. The same can be said for dual-purpose hemp crops regarding weed control.

The idea is to have a weed-free seedbed at the time of planting, so the hemp crop can eventually out-compete the weeds that will also emerge. It is important to plant hemp as soon as the soil has dried enough for planting and after the last intended seedbed preparation. The reason for that is weeds will also start to germinate as soon as the soil is firmed for planting. One does not want to give weed seedlings several days of lead time over the hemp seedlings.

Tine weeders are useful because hemp seedlings that are close to emergence will be anchored in the soil and are less likely to be torn out by the tine weeder. Weed seedlings that have germinated a day or two after hemp germination will more likely be exposed to the surface air and dessicate because they are not securely anchored in the soil. A tine weeding pass may be the last chance in narrow six- to seven-inch rows of fiber hemp to mechanically kill weeds.

European narrow 12-inch planters adapted for hemp may be desirable for seed/grain production. Their disadvantage is the longer time required to canopy closure, especially because hemp seed/grain varieties tend to grow slower than fiber varieties.

We were able to successfully cultivate 12-inch rows of hemp in 2019. A local John Deere dealer, Hillsboro Equipment Inc., fabricated a 12-row 12-inch spacing cultivator out of two used six-row 30-inch spacing cultivators. The use of GPS guidance on the tractor allowed us to successfully cultivate weeds between the narrow rows of hemp seedlings without compromising stand quality.

Timely cultivation allows the slower-growing wider row seed/grain crop to compete with weeds. Cultivation can be started at an early hemp-growth stage, which is two inches in height. Ground speed of the cultivator tractor would be about two miles per hour at that crop height. Multiple cultivations one week apart would maintain weed control between the rows. The third cultivation could occur when the hemp crop is about 12 to 18 inches tall at a ground speed of five miles per hour. That speed would allow some soil to flow within the hemp rows and bury some smaller weeds on the row.

Cannabinoid testing

Grain/seed varieties are historically low in grain CBD/THC content. It is extremely unlikely that hemp seed would exceed 0.3% THC. Samples of seed can be taken by the time the crop has set 50% of its seed (mid-September). Those samples can be sent or delivered to an approved lab for CBD/THC analysis, so hemp-grain potency will be known before harvest.

Fiber varieties are harvested before little or, sometimes any, seed is formed. Because fiber varieties will be self-pollinating, it is important to take samples for their CBD/THC potency when flowers start to form. Fiber varieties are also unlikely to exceed the 0.3% THC threshold because they are genetically low in both CBD and THC.

Harvest

Hemp grown for seed/grain should be harvested when shattering begins. The rest of the plant will still be green, and about 70% of the seed will be mature. Moisture content of the grain will likely be 22% to 30%. Grain combines can be used for grain/seed harvest with suggested settings similar to those used for grain sorghum. The long green stems of hemp can challenge combine harvest. Some growers have placed PVP pipes around moving parts to reduce wrapping.

As with any grain/seed crop, the proper harvesting, processing, transportation, and storage are critical to prevent spoilage and ensure the highest value for the harvested grain. Hemp grain is thin-walled and fragile, so it requires gentle handling. Grain/seed should be dried immediately after harvest to about 10% moisture or less.

Fiber-style hemp is cut with a sickle mower between early flowering and the onset of seed formation, when the lower leaves of the female plants begin to yellow. The cut plants are left in the field for as long as five weeks to allow retting to occur. Retting is a decomposition process that breaks the bonds between the outer long bast fibers and the inner short hurd fibers. The hemp is then raked into windrows two to three times to allow more drying and to remove as many leaves as possible.

The dried windrows are baled and shipped for processing, which separates the bast and hurd fibers. Bast-fiber concentration is highest in the bark of the stem, while high lignin but shorter hurd fibers dominate the core. Wider diameter stems are preferred. Planters that can singulate seed within a row are likely to produce a hemp crop with more uniform stem size.

This type of fiber hemp can also be cut with a kemper-style forage chopper head, then laid directly into a windrow without further chopping. Some self-propelled units are coming to market that have the ability to harvest the top portion of the plant and side-discharge the material onto trail wagons and cut the lower fiber portion with a kemper head — all in one pass.

History of crop rotations

Because hemp is adept at extracting soil contaminants such as pesticide residues and heavy metals, selected fields should have a crop history of winter rye, winter wheat, brassica cover crops, pasture grass sods, and/or alfalfa/grass sods immediately prior to growing a hemp crop. That's because those prior crops are less likely to have had pesticides applied to them.

Hemp can be grown successfully for consecutive years on a field. However, the longer one plants a field to the same crop, the greater the likelihood that crop will be affected by diseases, insects, and/or weeds that favor that crop. Hemp is no different than other crops in that regard.

An ideal rotation for growing hemp in the same field is planting it one out of every three years. If none of the previously mentioned cover crops are available for hemp production, prior crops such as those from the U.S. Department of Agriculture's Farm Service Agency Conservation Reserve Program (CRP), organically grown soybeans, corn, sweet corn, and vegetables are acceptable.

NOTE: A special caution is recommended regarding soybeans. Both hemp and soybeans are susceptible to a disease known as white mold. If a preceding crop of soybeans was infected with white mold, that disease leaves a structure behind in the soil known as sclerotia. Those sclerotia would have a high likelihood of germinating for as long as one to five years later and cause white mold in a susceptible crop such as hemp.

Crop rotation not only helps to avoid and minimize pests in current crops, but also it aids in preventing future pests from infecting crops.

Hemp is a dicot plant, which means the seed splits into two cotyledons as it emerges from the soil. Because of that, hemp should be planted following a monocot plant, such as grass, rye, wheat, barley, oats, triticale, and corn. Insect and disease pests are less likely to follow from monocot to dicot. Other dicot crops besides hemp include soybeans, clovers, birdsfoot trefoil, alfalfa, vegetables, lupines, edible beans, and canning peas.

Alfalfa is a dicot like hemp, and it provides enough nitrogen to grow a succeeding hemp crop in the same field. This fact is enticing to hemp growers, but caution is recommended because it is not a monocot plant. White mold in alfalfa is rare in northern latitudes, so that devastating disease of hemp is not likely to occur following a crop of alfalfa in the United States and Canada.

Preplant testing

Hemp uses the process of bioremediation to absorb pesticide and heavy metal contaminants in soils. As an example, Italian farmers grow hemp to clean downwind deposits of such contaminates from power plants. After a couple of years of hemp production, those fields can then again be used for edible vegetable production. However, when growing a production hemp crop, we are trying to find fields that exhibit already low levels of such contaminants. We do not want the production hemp crop to be unusable because of the contaminants it has absorbed.

In order to grow feminized hemp for flower and/or edible grain production, one should know the existing level of nutrients and residual pesticides in the soil and water proposed for hemp cultivation. Each state has accepted and regulated procedures for soil analysis for nutrients, which include pH, organic matter (OM), nitrogen (N), phosphorus (P), potassium (K), sulfur (S), calcium (Ca), magnesium (Mg), manganese (Mn), zinc (Zn), copper (Cu), iron (Fe), and chloride (CL). Laboratory names and contacts are available online for performing such soil analyses. Use a lab that has accreditation for your state.

Many such soil-testing labs also analyze domestic and irrigation well water that may be used for hemp production. The labs also may analyze soil and water for residual pesticide levels. An example of a regional laboratory that analyzes both soil and well water for nutrients and residual pesticides would be Midwest Laboratories. Eurofins Lancaster Laboratories is an example of a lab that performs soil and water analysis for residual pesticides.

Soil and water sampling and analysis should be done in the fall or spring prior to planting hemp.

Use proper soil-sampling techniques.

1. Use a stainless steel soil probe or auger to remove a uniform soil core to a depth of eight inches, and deposit the core into a plastic pail.
2. Collect at least two cores of soil from each acre of soil into the same pail.
3. Each sample submitted to the lab must contain at least six eight-inch cores. If your hemp-production area is less than three acres in size, you still must have a minimum of six cores per sample.

4. Pour the total of at least six cores into a soil-sample bag, which you can obtain (usually free of charge) from your lab of choice. Box the soil samples, and ship them off to the lab immediately. Try to ship samples between Monday and Wednesday, so the samples arrive at the lab by Friday.

Water samples can be obtained from domestic wells and irrigation wells used as the source for watering a hemp crop. Allow the well to pump for 10 minutes before collecting one quart of water into a one-quart plastic jar. Seal the lid and ship off to the lab early in the week.

A battery of four heavy metals (arsenic (As), cadmium (Cd), lead (Pb), and mercury (Hg)) can be obtained as a package analysis for soil and water from labs involved in heavy metals analysis. Additionally, an assay of 60 commonly used pesticides can be obtained from those same labs. If a grower suspects residual herbicide may be present in or around a selected field, be sure the battery of pesticide to be analyzed contains the suspected herbicide.

A list of the most common herbicides used in field corn and soybean production in the United States in the past 60 years includes but is not limited to: atrazine (AAtrex), simazine (Princep), metribuzin (Sencor), alachlor (Lasso), metolachlor (Dual), acetochlor (Surpass or Harness), glyphosate (Round Up), glufosinate (Liberty), paraquat (Gramaxone), linuron (Lorox), 2-4D (Ester and Amine formulations), and sulfentrazone (Authority).

Soil types

Hemp does not tolerate wet poorly-drained soils. Such soils are low in dissolved oxygen and often promote loss of plant-available nitrogen due to a process known as denitrification. In that process, plant-available nitrate nitrogen (NO_3-N) is converted to gaseous forms of nitrogen (NO, N_2O) by bacteria in the soil that are seeking oxygen. Because wet water-logged soils are low or devoid of dissolved oxygen, those bacteria utilize the oxygen in the NO_3-N molecule, which causes the conversion of the nitrogen to nitrous oxides. Those oxides bubble up through the soil water to the surface, where they are lost to the atmosphere.

In addition to loss of nitrogen, wet soils harbor root-rotting diseases that stop or severely limit hemp growth, so it becomes a straightforward recommendation to avoid hemp cultivation on poorly-drained soils.

Sandy soils also can be a poor choice for hemp cultivation. Such soils are prone to drought unless irrigation is available. Other limiting factors for sandy (even irrigated sandy) soils are that they are low in nutrient retention. Those soils are subject to downward movement in the soil profile to below the root zone, which is called leaching. They have difficulty retaining anionic nutrients such as nitrate nitrogen (NO_3-N), sulfate sulfur (SO_4-S), and chloride (Cl). In the past 20 years, it has become evident that even a cationic nutrient such as potassium (K) can leach within sandy soils. If one has irrigation available and is skilled at spoon-feeding nutrients to a crop of hemp throughout the growing season, then sandy soils can be a good soil type on which to grow that crop.

Soil types that have natural internal drainage, hold adequate available soil water (at least two inches per foot of depth), have good nutrient-holding ability (CEC of at least 4), and resist compaction are ideal for hemp production. Such soils always have the word "loam" as part of their soil name.

Refer to Figure 1.1, which depicts the universal soil textural triangle. In the middle of the triangle are the loam soil types known as loam, silt loam, sandy loam, clay loam, silty clay loam, and sandy clay loam. The textural triangle is used to determine how much of each of the three mineral components (sand, silt, and clay) are present in any given soil type. All three components added together always equal 100.

Clay soils often have a limit to their workability. The soil moisture needs to be just right (about 15% to 25% field moisture capacity) to till such soils. Those soils are susceptible to compaction by wheel traffic. Additions of liberal amounts (10 to 30 tons per acre) of solid manures are often useful in amending clay soils for hemp production as well as being useful for all soil types.

If one is unsure of what soil is in a field, the online "Web Soil Survey" tool is available from the U.S. Department of Agriculture's Natural Resources Conservation Service at no charge. After starting the tool by clicking on the green button, navigate to your state and county on the left side dropdown options. Once those selections are made, click on the "View" button next to county and state. Expand the map until you can pick out the field location (can also be more precise by using lat/long location if available). Then use the "Area of interest" tool to outline the selected field. Once the AOI is drawn, the "Soil Map" selection button will bring in the soil type for the selected area. An informational guide to the left of the map will describe the soil types that exist at the -selected site for your reference.

EPILOGUE

If you have made the decision to grow acres of cannabis, you are likely doing so believing that the world of tomorrow will not be the same as the world of yesterday. It's thought that the world of tomorrow will need crops like fiber hemp to use as sustainable raw materials. Those raw materials are necessary to produce consumer goods that have less detrimental impact on our globe. The consumer base in the world of tomorrow will prefer to purchase pain-relieving products that are non-addictive and come from natural processes. An expanding population consuming larger quantities of materials and goods will need to do so without jeopardizing future generations.

Growing cannabis can be a risky, challenging, and stressful undertaking. We certainly have a first-hand understanding of all of those aspects. But, we have also found a strong sense of satisfaction and accomplishment that follows a successful grow. We have asked many other growers who took on this project themselves with many differing focuses, sizes, and levels of success what they enjoyed most about the process. Most growers see the crop with the same eyes we do.

When you walk through, or stand in the middle of, a mostly mature hemp field during the growing season you will be overloaded by many senses. First and foremost is smell. The smell of growing cannabis is strong, some say intoxicating. One of our fields was located adjacent to a state highway and many locals commented on the aroma as they drove by with their windows down. Many commented on how "they never would have thought they would have lived to see the day…"

Change is constant. A common saying is, "Nothing is certain except death and taxes." Others have amended that, "Nothing is certain except death, taxes, and change." We walk this path daily. Many may not see the path that lies ahead, but it is likely not the same path they traveled on to get to where they are today. That is change. Consumer preferences are changing, the climate is changing, and most modern industries are changing.

Agriculture is no different.

We are no longer breaking sod ground with a team of oxen, one single plow shank, and the sweat off our backs. Both horsepower and acre units are actually defined directly from the days of old. From that, we changed to the days of large internal combustion engines exposing large swaths of earth annually from large-scale row-crop production. Today we have tractors that automatically steer themselves based on satellite reference points. We have planters that precisely place seeds with sub-inch accuracy over thousands of acres. We consider this an impressive feat that is continuously improved upon using technology.

The agriculture of tomorrow is slowing coming into focus. It's like looking through a dirty pair of glasses and squinting to see what is in front of you. Already there are completely autonomous machines with artificial intelligence that "weed" fields without the use of herbicides. Those machines apply electrical current to weeds to kill them but leave the crop unharmed. These machines identify plants that are not the same as the production crop, eliminate them in one pass, and continue driving themselves across the field. Deere & Co. has already revealed its electric prototype tractor that does not have any need for a human driver.

An Australian company has created eShepherd. Their product is transmitter collar units designed for livestock. The units are synchronized to satellites, base stations, and a laptop computer or phone using GIS software to track and move livestock. Science has proven that rotational grazing is better for the land itself, beneficial for livestock, and increases water quality. Internal paddock fencing for the farmer or rancher is costly, labor intensive, and time consuming to operate and install. Creating an invisible fence that livestock graze within and changing it daily with your computer alleviates these constraints. It will undoubtedly change the world.

The use of these technologies, along with many others will allow the producers of tomorrow to deliver high-quality nutritious foods to a growing consumer base without harm to the environment. In many ways, farming may lead the sustainable revolution. Already, there are companies offering to pay farmers on a per-acre basis to demonstrate their products' abilities in regards to carbon sequestering. The concept is storing excess carbon from the atmosphere to create a smaller footprint for tomorrow. Many other industries will have to follow suit in order to survive.

Many industrial hemp promoters reference the Henry Ford quote, "Why use up the forests which were centuries in the making and the mines which required ages to lay down, if we can get the equivalent of forest and mineral products in the annual growth of the hemp fields?"

We tend to agree with Henry on this one. Seventy years of a hiatus in the advancement of technologies pertaining to hemp have held its practical usage back. The brakes now have come off. The race is on to develop the specific consumer products from hemp raw materials that will be used well into the future.

Many will not finish this race.

Some failed prior to making the first step. Others will tire before the finish line. Those that succeed to cross that line will have forever changed the world as we know it.

Successfully cultivate cannabis, and it may very well be you who wins the race.

GLOSSARY

Alternative cannabinoids – there are more than 100 differing cannabinoid compounds that come from the cannabis plant. Alternative cannabinoids refer to those that are not CBD or THC, which are the two most widely known cannabinoids.

Bioremediation - the process of using microorganisms to treat polluted water, land, or waste. Cannabis as a bioremediator can pull contaminates from the soil and store them in its biomass, therefore remediating the soil itself.

CBD - stands for cannabidiol, which is a compound that grows naturally in the flowers of the cannabis plant. CBD is often used as a medicine for therapeutic purposes or treatments.

Certificates of Analysis - proof that a varietal has gone through a form of quality control to ensure it has certain specifications.

Certified Crop Advisors - agronomists who have become certified through the American Society of Agronomy. The certification entails having the "commitment, education, expertise, and experience to make a difference in a client's business," according to www.certifiedcropadviser.org.

Conservation Reserve Program - is a program through the U.S. Department of Agriculture's Farm Service Agency. It is a cost-sharing and rental payment program used by farmers.

Crop plan - a farmer or grower's way of staying on track to ensure maximum production and profitability from a designated acreage

Cultivar - a plant that has been established using selective breeding. A plant variety occurs naturally, while a cultivar has been altered by humans.

Feminized seed - this seed is bred specifically to only contain female plants. However, there is almost always a small percentage that remains male.

Genome - the genetic makeup of an organism contains genomes, which contain DNA.

Organic Materials Review Institute – A third party organization that ensures products that are intended for USDA certified organic usage meet certain requirements.

Photoperiod - the time during the day that the plant receives sunlight.

Standard Operating Procedures - businesses often create sets of rules or guidelines specific to the organization. Farms have Standard Operating Procedures in order to maintain consistency and routines.

THC - stands for tetrahydrocannabinol, which is a compound that grows naturally in the cannabis plant. THC is the active ingredient that produces a "high."

Throughput - this refers to an amount of product going through a system.

Varietal - another way of saying variety. Different varietals can form a variety.

INDUSTRIAL HEMP VENDORS - WISCONSIN

Fertilizers
- Agri-Energy Resources – Gary Campbell
  - Supplier of OMRI-approved fertilizers (Chilean nitrate) and OMRI-approved foliar products
- Allied Cooperative
  - Supplier of OMRI-approved fertilizers including Chilean nitrate
- BFG Supply – Jason Jacobs
  - Supplier of greenhouses, greenhouse products, and OMRI-approved foliar products
- Dramm Corporation
  - Supplier of emulsified fish products
- Midwest Bio-AG
  - Supplier of OMRI-approved potassium sulfate
- Purple Cow Organics
  - Supplier of OMRI-approved organic medium
- QLF
  - Supplier of liquid sugars (Terrafed), Kelpak
- United Cooperative
  - Supplier of OMRI-approved fertilizers including chicken litter

Pesticides – OMRI-approved
- Azadirachtin – Many products and manufacturers available online
- Beauveria bassiana – Many products and manufacturers available online
- Bacillus thuringiensis (Bt) - Many products and manufacturers available online

Beneficials

Beneficial Insects – Arbico Organics Website: https://www.arbico-organics.com/category/beneficial-insects-organisms

Organic Certifiers

https://www.ams.usda.gov/services/organic-certification/certifying-agents

Driftless Extracts 2020 Feminized Hemp Application Recommendations

Stage	Product	Amount	Notes
Pre-seed (Greenhouse)			
Treat Seed	Kelpak	8oz/100 lbs seed	Spray over seed, let dry, seed into media
Fertigate	Kelpak	1% Dilution	Inject into water (pulse) at a quart/25 gal
	Zinc Sulfate	1% Dilution	
Post Germination (greenhouse)			
2-3 leaf stage	Kelpack	1% Dilution	Inject into water
	TerraFed	0.5% Dilution	
5-6 leaf stage	Liquid Fish	0.5% Dilution	Inject into water
Prior to transplant	Kelpack	1% Dilution	Inject into water
	Terrafed	0.5% Dilution	
	Liquid Fish	0.5% Dilution	
Field Prep/Transplant (outdoor)			
Prior to raising beds	Terrafed	5lb/A	For prior crop breakdown
Pre-Plant Fertilizer Banded - If possible	Potash	200#/A	*100 #/A K-MAG NATURAL*
	Chilean Nitrate	200#/A	~~[struck through]~~
Raise beds	If necessary	48"width X 12" Height	
Planter band	TerraFed	2 gal/A	Rate of product should be banded w/4-6 inch treatment
	Serenade Opti or Trichoderma	Labeled Rate	
	Active 5-10-10	8lb/A	
	Beauvaria Bassiana	Labeled rate	
	Zinc Sulfate	1 gal/A	
	Mn Sulfate	1 gal/A	
	Boron	1 qt/A	
Cover crop	Winter Rye	¾ bu/A	Broadcast seed
After Transplant (Outdoor)			
Water	Irrigation	6 gal/plant/week	
5 days after transplant	TerraFed	2 gal/A	Foliar Application #1
	Fish	2 gal/A	
	MN S04	1 gal/A	
2x transplant height (18" Plant height)	Kelpak	1.5pt/A	Foliar Application #2
	Liquid Fish	2.5lb/100gal	
	Potassium Bicarbonate	1-2#/A (fungicide)	
	Zinc Sulfate	0.5pt	
	Mn Sulfate	1/8 pt	
Last week of ~~[struck]~~ *JULY*	Chilean Nitrate	~~[struck]~~ *200#/A*	Pre-bud *200#/A POTASSIUM SULFATE*
Insecticide	Azadirachtin	Labeled Rate	If Necessary
Fungicide - Pre Bud	Copper Fungicide	Labeled Rate	Apply Pre-Bud using Application #2 Products
Fungicide - Full Flower	Serenade Opti	Labeled Rate	If Necessary

Disclaimer: Applications should follow all state and federal laws governing usage. Apply using safe application procedures and required documentation. These recommendations do not ensure successful cultivation of hemp as a cash crop. Follow all product labels including PPE and REI implementations. Tank mixes should be properly monitored to ensure crop safety.

Indica Dominant

Sativa Dominant

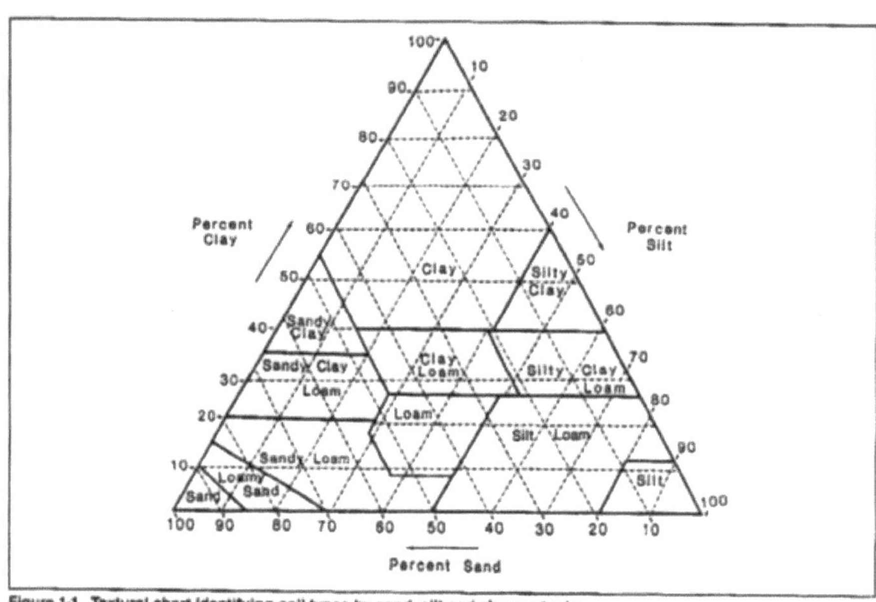

Figure 1-1. Textural chart identifying soil types by sand, silt and clay contents.

WORKS CITED

Schulte, E. E., & Walsh, L. M. (2005). *Management of Wisconsin Soils* (5th ed., Vol. A3588). University of Wisconsin-Extension.

Wright, A. H., (1918) *Wisconsin's Hemp Industry*. (Bulletin 293). Agricultural Experiment Station of the University of Wisconsin, Madison.

Huenfeld, Donald G. (2019) QLF Agronomy. *email*. 13 April